P9-DER-010

Planet Earth

ED RILEY
SANDRA FORTY

Published by World Publications Group, Inc.
140 Laurel Street
East Bridgewater, MA 02333
www.wrldpub.net

First produced in 2008 by BlueRed Press Ltd.
6 South Street, Totnes, Devon, TQ9 5DZ
www.blueredpress.com

Designed by Jon Morgan at J2 Design

ISBN–10: 1-57215-384-9
ISBN–13: 978-1-57215-384-4

Printed and bound in China

Contents

6 Introduction

8 Mountains

36 Oceans

58 Deserts

76 Woodlands

98 Iceworlds

122 Rainforests

146 Great Plains

168 Wetlands & Freshwater

189 Credits 190 Index

Introduction

The planet that we live on is estimated to be somewhere in the region of 4.5 billion years old and we humans have only been around for the merest fraction of that time—some 250,000 years.

Over the millennia the shape and structure of the Earth's landmasses has undergone a vast amount of change—from the C-shaped supercontinent, Pangaea, that existed some 270 million years ago and contained virtually all of the world's landmasses (as well as occupying almost a third of the planet's surface surrounded by a global ocean called Panthalassa) to the seven continents and five oceans that we recognize today.

When the evolution of life on Earth began is still a matter of scientific debate, but the earliest beginnings of plants and animals that would be recognizable to us occurred some 550 million years ago with marine invertebrates, such as shell-making ammonites. Fish, then amphibians, reptiles, birds, mammals, and—lastly—human beings evolved from these marine invertebrates.

Up until 200 years ago virtually all the physical changes that occurred in the world were natural, from the forming of mountain ranges caused by the uplift of tectonic plates to the gouging of valleys by the inexorable movement of glaciers. But the industrial revolution of the 18th century and the rapid advances in technology that came with it led to humans beginning to drastically alter the Earth's landscape—from widespread destruction of the world's forests to the loss of many animal species through over-fishing and hunting.

What follows is a visual celebration of some of the most enduring of nature's wonders as well as some of those areas and species that are, even now, most in danger of being lost forever.

Fortunately, in more recent times there has been a far greater awareness of the importance of the environment and our impact upon it and global efforts are being made to attempt to redress the balance.

Mountains

Mountains make up roughly one-fifth of the world's landscape and are among the most awe-inspiring of nature's sights, whether a lone peak towering over the surrounding area, such as the Matterhorn on the border between Switzerland and Italy, or an entire mountain chain, such as the Andes, which dominates the western coastline of South America.

What Is A Mountain?

While examples such as the Matterhorn and the Andes are self-evidently mountains there is no universally agreed definition of what constitutes a mountain. The Collins English Dictionary defines a mountain as "a natural upward projection of the Earth's surface, higher and steeper than a hill and often having a rocky summit." Simple enough, but the problem is that one man's mountain is another man's hill. The Watchung Mountains—a group of three ridges in northern New Jersey—range from just 400 ft (122 m) to 500 ft (152 m) high, and compared to Himalayan peaks that reach over 8,000ft. (2,438 m) barely register as bumps in the landscape. The U.S. Board on Geographic Names sums up the problem thus "…the difference between a hill and a mountain in the U.S. was 1,000 feet of local relief, but even this was abandoned in the early 1970s. Broad agreement on such questions is essentially impossible, which is why there are no official feature classification standards."

How Mountains Are Formed

Planet Earth is covered by a "skin" of rock, known as the lithosphere, which is made up of fragmented segments called tectonic plates. At this point in time there are seven major and a number of minor tectonic plates. Of the major plates six are continental: the African Plate, the Antarctic Plate, the Australian Plate, the Eurasian Plate, the North American Plate, and the South American Plate. The seventh is the Pacific Plate, which is oceanic.

The movement of these tectonic plates causes the creation of the majority of mountains on the Earth's surface. Compressional forces form "fold" mountains, such as the European Alps and the Himalayas in Asia. Where two tectonic plates collide they force sedimentary rock upward and the rock folds back on itself to create a ridged pattern of peaks and troughs.

"Fault block" mountains, such as the Teton Range in Wyoming and the Sierra Nevada mountains in California, are formed when faults or cracks in the

Previous page: Mont Blanc Massif is a range in the western part of the Alps. It is named after Mont Blanc, the highest summit of the Alps at 15,774 ft (4,808 m). Spanning France, Italy, and Switzerland, it is sometimes considered to be part of the Pennine Alps or the Graian Alps.

Left: The Ouzoud Waterfalls are 330 ft (110 m) high and are located in the Middle Atlas mountains, in the Grand Atlas village of Tanaghmeilt, in the province of Azilal, 93 miles (150 km) north-east of Marakech, in Morocco. They are the most visited site in the region. Ouzoud is the Berber word for "olive," and refers to the nearby olive trees.

Above right: Sangay is a constantly active stratovolcano, located east of the Andean crest. Its steep cone shape and glacier-covered top towers majestically 17,159 ft (5,230 m) above the Amazonian rainforest. It is one of the highest volcanoes in the world and one of Ecuador's most active ones. Sangay National Park is part of the World Heritage Forest program.

Earth's surface force large blocks of rock upward. Rather than folding over, like fold mountains, these blocks of rock become stacked in parallel lines.

"Dome" mountains, such as those in the Black Hills of South Dakota and the Adirondack Mountains of New York, are formed when molten rock (magma) pushes up through the Earth's crust thrusting the rock above it upward, creating a circular dome on the surface. Peaks and valleys are formed as the rock erodes over time.

Volcanic mountains, such as Mount St. Helens in Washington and Mount Vesuvius in Italy, occur when molten rock erupts through a fault in the Earth's surface. As the molten rock solidifies around the fault it forms the familiar cone-shaped mountain with a central crater.

Plateau Mountains, such as the Catskills in New York State and the Ozark Mountains in Arkansas and Missouri, are formed by erosion. Because different types of rock are eroded at a different rate areas of hard rock surrounded by softer, more easily eroded, rock will eventually be left standing high above the surrounding area.

The sheer size and imposing presence of most mountains lends them an air of permanence, yet even these giants of the landscape have a lifecycle that will eventually see them disappear, worn away by the forces of erosion.

North America – The Rockies

The Rockies are the major mountain range in North America. Made up of over one hundred individual ranges, the Rockies stretch from northern Alberta and British Columbia in Canada down to New Mexico in the United States. Covering a distance of around 3,000 miles (4,800 km), they are at points over 300 miles (480 km) wide.

Although the many ranges within the Rockies differ greatly in terms of their age and despite the vast area that they encompass, they share many common characteristics such as high peaks—many standing over 13,000 ft (4,000 m)—rich mineral resources, and dramatic scenery caused by historic volcanic and glacial activity.

The weather in the Rockies is characterized by cold winters and cool summers. However, as the range stretches from the tip of the subtropical zone in New Mexico to the Arctic in Canada the extremes of the climate become more notable. The further north you travel the shorter and cooler the summers become and the longer and more severe the winters. While in the Southern Rockies many of the highest peaks are snow-capped for some of the year in the northern reaches there are year-round glaciers in a number of the higher valleys. The climate in the southern areas of the Rockies also tends to be

Left: The European lynx (*Lynx Lynx*) is a medium-sized cat native to European and Siberian mountain forests. They are primarily nocturnal and live solitarily as an adult. Once quite common across all of Europe, by the middle of the 20th century they had become extinct in most countries of Central and Western Europe though there have recently been successful attempts to reintroduce lynx. This female was photographed in a winter birch forest.

Right: Mongolia is the world's nineteenth-largest country, with the Gobi desert to the south and cold, mountainous regions to the north and west. It has an extreme continental climate with long, cold winters and short summers. A "yurt," pictured here in the Altai Mountains, is the traditional dwelling of the nomadic people in Mongolia. Easy to collapse and assemble again it can be transported on no more than three animals.

considerably drier than in the north with the majority of rain and snowfall occurring during the winter, whereas in the north there is year-round precipitation.

Despite the seemingly harsh environment, wildlife in the Rockies flourishes and is both plentiful and diverse. Bighorn sheep and mountain goats thrive in the rugged peaks during the milder summer months and migrate down to the lower slopes during the winter. Birds of prey such as bald eagles, golden eagles, ospreys, and peregrine falcons can be found nesting throughout the range. Two species that came close to extinction can still be found in the Rockies: one of the largest herds of bison in the U.S. is in the Yellowstone National Park in Wyoming and wolves, though still rare, are slowly repopulating the Rockies.

Perhaps the animal most emblematic of the American mountain wilderness—the grizzly bear—is also to be found in the Rockies. Despite being thought of as carnivorous hunters, grizzlies feed on both plants and animals. They spend the summer months on the lower slopes fattening up for the winter when they retreat up the slopes to spend six months in near-hibernation living off the fat reserves they built up during the summer.

South America – The Andes

Situated on the west coast of South America, the Andes are both the longest mountain chain on Earth and one of the highest. Covering a distance of around 5,500 miles (8,900 km) they stretch from Colombia and Venezuela in the north down to the southernmost tip of Chile. Over fifty of the peaks in the Andes stand higher than 6,000 m (19,700 ft) above sea level. The Andes is also home to some of the world's highest active volcanoes, including Tupungato at 6,570 m (21,555 ft) in Chile and Cotopaxi, which rises to 5,897 m (19,342 ft) in Ecuador.

The Andes can be separated into three broad, geographic sections: the Southern Andes in Chile and western Argentina, the Central Andes in Chile, western Bolivia and Peru, and the Northern Andes in Ecuador, Colombia, and Venezuela. The climate varies greatly throughout these three regions; the north, being close to the equator, is

Above: A yak (*Bos grunniens*) pictured in the Himalayan snowcapped mountain landscape of Central Asia. Yaks are herd animals and usually form groups of between ten and thirty animals. Their habitat is treeless uplands, hills, mountains, and plateaux between 10,500 ft (3,200 m) and 18,000 ft (5,400 m). Both male and female have long shaggy hair to insulate them from the cold and at night and in snow storms they will huddle together, keeping calves in the center, pressed so close together the condensation from their breath rises into the air like a column of steam.

generally warmer and more humid with rainforests situated only a few miles from the snow-capped peak of Cotopaxi. The Southern Andes, at its extremity nearer to Antarctica than the equator, is much colder, while the Central Andes are dryer and milder than the other two regions. The dramatic difference in the climate from one end of the range to the other is perfectly demonstrated by the height of the tree line. Close to the Antarctic in the far south of Argentina the tree line does not rise much higher than sea level, whereas in more tropical areas, such as Bolivia, trees grow as high up as 4,000 m (13,120 ft).

While the rainforests that used to encircle large areas of the Northern Andes are greatly diminished they are still home to an abundance of fauna and flora, much of it unique to the area. The region is home to one of the largest flying birds on Earth, the Andean condor, whose wingspan can measure up to 3.2 m (4 ft), as well as the only bear to be found in South America, the spectacled bear (so called because of the white circles around their eyes that look like spectacles).

Europe – The Alps

The Alps are the highest and most densely populated mountain range in Europe. Although mainly situated in Switzerland and Austria, the Alps also pass through another seven countries: France, Italy, Germany, Slovenia, Croatia, Bosnia and Herzegovina, and Yugoslavia, covering a total area of around 80,000 square miles (207,000 square km). The highest peak in the range is Mont Blanc also known as "La Dame Blanche" (French for "the white lady"). Located on the French/Italian border the peak stands 4,808 m (15,774 ft) high.

The climate of the Alps is one of extremes. Due to its location in central Europe, the area is affected by weather fronts from all four points of the compass. From the north the area is buffeted by cold, polar winds; from the east the winds blow cold and dry in the winter and hot and dry in the summer; from the south and west comes milder, moister air from the Atlantic and the Mediterranean. Generally though the alpine valleys are warm and dry with the majority of precipitation falling above 5,000 ft (1,500 m) in the form of snow. This rough division can also be seen in the trees that grow in the Alps, the lower slopes are dominated by deciduous trees such as oak, beech, and poplar while higher up the slopes coniferous trees, such as the spruce, larch, and pine proliferate. At higher levels the trees give way to Alpine tundra that is home to a variety of plant life including rhododendrons, edelweiss, gentian, and dwarf shrubs. During the short Alpine

summer these plants give the area an enchanting, colorful appearance.

Although the higher reaches of the Alps are a harsh environment, some animals have adapted well enough to the conditions to thrive; the sure-footed Alpine ibex and squirrel-like marmot among them. Many areas of the Alps are now protected as parks and reserves in an attempt to preserve the wildlife that lives there.

Africa – East African Mountains

The East African mountain region encompasses parts of Kenya, Tanzania, Uganda, the Democratic Republic of the Congo (D.R.C.), Rwanda, and Burundi. Most of the mountains within this area are volcanic in origin with the most notable exception being the Rwenzori Range on the border between Uganda and the D.R.C. The highest mountains within the range stand over 6,000 ft (4,900 m) high and in spite of their proximity to the equator the peaks of the higher mountains are capped with ice.

The highest East African mountain—and the highest in Africa—is the dormant volcano Kilimanjaro in northeastern Tanzania. Kilimanjaro consists of three volcanic cones, Kibo, Mawenzi, and Shira, the highest of which—Kibo—is 19,341 ft (5,895 m) above sea level.

Despite being in an area that is mainly dry the mountains receive a high amount of rainfall—the most falling in the Ruwenzori, where an annual rainfall of over 70 inches (1,778 mm) is not uncommon at heights of over 15,000 ft (4,572 m).

The lower slopes of the mountains follow the expected pattern of grasslands at the lowest levels, leading to forestation at the mid-levels (including some giant trees, such as camphor, cedar, and the East African olive). Above the tree line can be found the Afro-Alpine zone, home to a number of plants that are unique to the area. Rhinoceroses, elephants, buffalo, antelope, and monkeys inhabit lower and mid-level areas, while antelopes and leopards can be found higher up.

Asia – The Himalayas

The Himalaya Range (from the Sanskrit for "abode of snow") separates the Indian subcontinent from the Tibetan Plateau. The sheer magnitude of the Himalayas is breathtaking; the range stretches for over 1,490 miles (2,400 km), spans almost 250 miles (400 k) at its widest point and is home to most of the highest mountains in the world, the highest of which—Mount Everest—is 29,035 ft (8,850 m) tall. To put this into perspective, consider that the highest mountain outside of Asia is Aconcagua in the Andes at 22,841 ft (6,962 m) while the Himalayas contain no less than fourteen mountains over 26,250 ft (8,000 m).

So vast are the Himalayas that they have an enormous impact on the climate of the surrounding areas. The mountains block both the passage of the southbound cold, dry Arctic winds into India and the northbound rain-bearing (monsoon) winds onto the Tibetan Plateau. This leads to the climate to the south of the mountains being warmer and wetter than might otherwise be expected while the climate to the north is more dry and arid than it would otherwise be.

Wildlife at the higher altitudes of the Himalayas is sparse but those species that do exist there include the snow leopard, the brown bear, the red panda—a species unique to the area—and the Tibetan yak.

Australasia – The Flinders Ranges

The Flinders Ranges is the largest mountain chain in South Australia, stretching northward from just south of Port Pirie for around 500 km (about 310 m) to just south west of Lake Callabonna (a dry salt lake). The highest point in the Flinders is Saint Mary Peak 3,825 ft (1,166 m). The mountain can be found in one of Australia's most famous landmarks, Wilpena Pound, which is a natural amphitheater of mountains in the middle of the Flinders Ranges National Park.

The Flinders Ranges contain some of the oldest mountains in Australia and in 1946 the fossilized remains of soft-bodied organisms from the late Precambrian period (over 500 million years ago) were discovered in the Ediacara Hills, north-west of Leigh Creek.

Top left: Photographed north of Jasper in May, Jasper National Park is the largest national park in the Canadian Rockies. It spans 4,200 miles (10,878 km) and is located in the province of Alberta. The park was declared a UNESCO World Heritage site in 1984 and contains mountain peaks, glaciers, lakes, waterfalls, canyons, and limestone caves.

Left: The Wairau River rises in New Zealand's Spenser Mountains and flows for 105 miles (169 km) between the St. Arnaud and Raglan ranges. Shown here running through the Lewis pass, North Canterbury, on the South Island the photograph shows slow moving sediment deposits.

Far Left: Half Dome as seen from Glacier Point in the Sierra Nevada mountain range. Located at an elevation of 7,214 ft (2,199 m) the point offers a superb view of Yosemite National Park. The granite crest of Half Dome rises more than 4,737 ft (1,444 m) above the valley floor.

Above: Grand Teton National Park lies in northwestern Wyoming, just to the south of Yellowstone National Park, and is a spectacular glaciated mountain region. The steep, rugged mountains give way to the morainic landscape of the valley, dotted with glacial lakes as pictured here. Within the Teton Range the tallest peak is Grand Teton at 13,770 feet (4,197 m).

Left: Designated a World Heritage site in 1978, Yellowstone Park includes the greatest concentration of geothermal features in the world. Morning Glory Pool was named in 1880 for its remarkable likeness to the morning glory flower. The distinct color of the pool is due to bacteria. On rare occasions, usually following nearby seismic activity, the pool erupts as a geyser.

Previous page: Spirit Island, in the middle of Lake Maligne in the Jasper National Park can only be reached by boat. The unspoiled beauty of this area makes it one of the most stunning and popular views of the Canadian Rockies.

Top: Dall sheep (*Ovis dalli*) can be found in the mountainous regions to the northwest of North America and are sure-footed climbers. They inhabit a combination of open alpine ridges, high meadows, and steep slopes and stay close to extremely rugged ground in order to escape predators that cannot travel quickly on such terrain. Their primary predators are wolves, coyotes, black bears and grizzly bears. This young Dall sheep was photographed in Brooks Range, Alaska.

Above: A hoary marmot (*Marmota caligata*) looks over the Exit Glacier in Kenai Fjords National Park, Alaska. This species primarily lives near the tree lines on slopes with grasses and herbs to eat and rocky areas for cover. It has a high-pitched warning whistle used to alert other marmots to possible danger. The word "hoary" refers to the silver-grey fur on the species' shoulders and upper back.

Right: These brown bear (*Ursus arctos*) were photographed at the Katmai National Park on the Alaskan peninsula. The species feeds on vegetable matter including roots and fungi, while fish is its main source of meat. The brown bear will put on up to 400 pounds (180 kg) of fat during the summer months, which it needs to make it through the winter. Mostly solitary, they may gather in large numbers to feed and form social hierarchies based on age and size.

Above: To the north of Puerto Natales are some of the world's most spectacular natural reserves, including the Torres del Paine National Park, which covers over 530,000 acres.

Right: At an average altitude of approximately 11,000 ft (4,000m) the Altiplano (which translates from Spanish as "high plain") is a vast expanse of high plateau in the central Andes. The region boasts crystal clear rivers, blue and emerald lakes, cactus forests, and abdundant wildlife and is surrounded by the highest volcanoes in the world.

Far right: Machu Picchu or "Old Mountain," is a pre-Columbian Inca site located 7,875 ft (2,400 m) above sea level in Peru. The city sits in a saddle between two mountains above the Urubamba Valley, with a commanding view down two valleys and a nearly impassable mountain at its back. It was declared a World Heritage site in 1983 and in July 2007 it was voted one of the New Seven Wonders of the World. However, in 2008 it was added to the watch list of the 100 most endangered sites as a result of the impact of tourism.

Above: Pictured here in the Cairngorms National Park, Scotland, the pine marten (*Martes martes*) is part of the weasel family and native to Northern Europe. Their habitats are usually well-wooded areas and they make their dens in hollow trees or scrub-covered fields. Pine martens are mainly active at night and dusk.

Main image: Tryfan Mountain in Snowdonia forms part of the Glyderau group and at 3,002 ft (915 m) is one of the highest in Wales. The Glyderau stretch from Mynydd Llanegai to Capel Curig, and include five of Wales's summits over 3,000 feet. Tryfan is considered one of the finest mountains in Wales and one of the few on the British mainland that requires scrambling to reach the summit.

Top: A herd of reindeer (*Rangifer tarandus*) grazing on tundra vegetation in the Caringorns National Park, Scotland. Also known as caribou when wild in North America, they are an Arctic and Subarctic-dwelling deer. Large populations of wild reindeer are still to be found in Siberia, Greenland, Alaska, and Canada. The last wild reindeer in Europe are found in portions of southern Norway.

Above: Golden eagles (*Aquila chrysaetos*) photographed in late May in Scotland. Usually mating for life, the eagles will build several eyries and use them alternately. Old eyries may be 6 ft (2 m) in diameter and 3 ft (1 m) in height, as eagles repair their nests whenever necessary. Mainly mountain dwellers due to habitat destruction, the goden eagles of Sweden and Denmark have now started to breed in lowland areas again.

Left: Fee Glacier above Saas Fee in the Valais region of Switzerland, in the Alps. A 100 years ago this glacier extended down the mountain to the edge of the village of Saas Fee, but has now retreated substantially. Over the five year period from 1995 to 2000, 103 of 110 glaciers examined in Switzerland, 95 of 99 glaciers in Austria, all 69 glaciers in Italy, and all six glaciers in France were in retreat.

Above: Above 11,500 ft (3,500 m) is the Afro-alpine zone, a moorland characterized by tussock grasses, senecios, and lobelias. In this photograph a giant senecios towers over the surrounding rocky moor, their thick stems servicing as reservoirs for precious water.

Below: Rock dassie or hyrax, (*Procavia capensis*) inhabit rocky terrain across sub-Saharan Africa. They live in small family groups dominated by a single male. They are well-furred rotund creatures and from a distance could be mistaken for a rabbit or guinea pig.

Above: Young mountain gorilla (*Gorilla beringei*) at Volcanoes National Park, Rwanda. The Mountain Gorilla lives within vegetation at altitudes from 7,300 ft (2,225 m) to 14,000 ft (4,267 m) and their mountain homes are often cloudy, misty, and cold. Today the species is threatened by poaching, loss of habitat, and disease and is now classed as critically endangered.

Previous page: Elephants in Amboseli National Park, Kenya with snow-capped Mount Kilimanjaro rising in the background. Amboseli, meaning "salty dust," is indeed dusty, mostly due to its proximity to Mt. Kilimanjaro. It is nevertheless lush in places because of melting snow that feeds the springs, swamps, and marshes. The park is famous for being the best place in Africa to get close to free-ranging elephants.

Above: The Bactrian camel (*Camelus ferus*) is native to the steppes of north eastern Asia and is one of the two surviving species of camel. The Bactrian has two humps on its back in contrast to the single-humped dromedary camel. Seen here against the Altai Mountains, Mongolia, it is endangered in the wild and the population is decreasing.

Left: Mount Elbrus is the highest peak in Europe. It is part of the Central Caucasus and its summit is capped in ice year round. There are countless glaciers that sprawl from its slopes and all told the mountain and its glaciers cover 56 square miles (145 square kms).

Below: The Cacusus Mountain range stretches for 550 miles (885 km) between the Black Sea and the Caspian Sea, forming a natural southern barrier between Asia and Europe. Though comparable to the European Alps, the Caucasus are considerably higher, averaging 6,000 feet (1,828 m) to 9,000 feet (2,743 m) and rising to 18,481 feet (5,633 m) in Mount Elbrus.

Above: Mount Everest is the highest mountain on Earth. Measured by the height of its summit above sea level, it rises to 29,029 feet (8,848 m) in the Himalaya range of High Asia on the border between Sagarmatha Zone, Nepal, and Tibet, China.

Far left: Nun Kun Massif comprises a pair of Himalayan peaks, Nun at 23,409 ft (7,125 m) and its neighbor Kun at 23,218ft (7,077 m). Nun is the highest peak in the range and lies on the Indian side. It is seen here from the Suru Valley, Western Ladakh.

Left: The Himalayas at sunrise.

Main image: A photograph of the Tengger massif in Java, Indonesia, showing the volcanoes Mt. Bromo (large crater, smoking) and Mt. Semeru (background, smoking). Fog surrounds the peaks, covering a plain of fine volcanic ash.

Far left: An aerial view of the Heysen range in the Flinders Ranges, South Australia. The range extends some 500 miles (800 km) northward from near Port Pirie to Lake Callabonna and comprises some of the oldest and most rugged mountains in Australia.

Center: This alpine tarn at Key Summit sits at 3,011 feet (918 m) and is accessed from the Routeburn Track, in the Fiordland National Park, South Island, New Zealand. The Fiorldland National Park is a major part of the Te Wahipounamu World Heritage site.

Below: Hooker Glacier, close to the slopes of Mount Cook (Aoraki) in the Southern Alps, is the source of the Hooker River formed by the meltwaters of the glacier. Mount Cook is the highest mountain in New Zealand at 12,316 ft (3,754 m).

Oceans

Previous page: The humpback whale (*Megaptera novaengliae*) is a baleen cetacean. This acrobatic animal was photographed in Hawaii breaching the water.

Left: A sea otter (*Enhydra lutris*) rests in the water at Prince William Sound, Alaska. Hunted extensively for their fur until 1911 the world population fell to below 2,000 at one time. Allthough their recovery is considered a success sea otthers are still at depressed levels and therefore remain classed as an endangered species.

Below right: This deepwater jellyfish normally lives at depths below 600 ft (200 m).

The famous "Blue Marble" photograph of the Earth taken by the crew of Apollo 17 in December 1972 demonstrates the predominance of water on the surface of the planet. In fact, just over 70%—an area of around 139,382,000 square miles (361 million square kilometers—of the Earth's surface is covered by water with 97% of this made up of salty oceans.

Despite being a continuous body, the salt water of the Earth is generally divided along geographical lines into five principal oceans—the Pacific Ocean, the Atlantic Ocean, the Arctic Ocean, the Indian Ocean, and the Southern Ocean as well as numerous smaller seas. The Southern Ocean was effectively "created" in the spring of 2000 when the International Hydrographic Organization decided to demarcate a fifth ocean comprised of southern sections of the Atlantic, Pacific, and Indian Oceans.

The Layers of the Ocean

Just as the Earth's atmosphere is divided into different layers, so the ocean is also made up of five layers, or zones, with distinctive characteristics.

The *Epipelagic Zone* (also known as the "sunlight zone") is the upper layer of the ocean that extends down from the surface to a depth of roughly 660 ft (200 m). It is here that most of the visible light in the ocean can be found. As well as light the sun provides heat to the Epipelagic Zone. Seasonal and geographic factors greatly affect the surface temperature of the sea, with temperatures ranging widely from as high as 97°F (36°C) in the Persian Gulf to as low as 28°F (-2°C) in the Arctic. Although life in one form or another exists in all five of the ocean layers it is here, thanks to the light and warmth of the suns rays, that the greatest abundance and variety can be found.

The *Mesopelagic Zone* (also known as the "twilight zone") extends from 660 ft (200 m) to a depth of

3,300 ft (1,000 m). Here the effects of sunlight become faint and there is far less light and warmth than in the Epipelagic Zone. Due to this lack of light it is here that we first see examples of bioluminescence, the emission of light by a living organism. Many of the fish at this level have larger, and in some cases upward facing, eyes to enable them to make the most of the limited light available. Temperature variations within the Mesopelagic Zone are even greater than in the Epipelagic Zone as the temperature decreases more rapidly with depth than anywhere else in the ocean.

The *Bathypelagic Zone* (also known as the "midnight zone") extends from 3,300 ft (1,000 m) to 13,100 ft (4,000 m). The suns rays do not penetrate here at all and the bioluminescence of the animals themselves is the only source of light. The temperature in the Bathypelagic Zone, unlike the two layers above, is relatively constant, rarely straying far from 39°F (4°C). At the deepest levels of the Bathypelagic Zone the pressure becomes extreme, reaching up to 5,850 pounds per square inch—over 800 times the pressure at the surface. Because of this vast difference in pressure if fish from the Bathypelagic Zone are brought to the surface the gases in their system expand, causing them to explode.

The *Abyssopelagic Zone* (also known as the "abyssal zone") extends from 13,100 ft (4,000 m) to 19,700 ft (6,000 m). Although the name "abyssopelagic" comes from the Greek word abyss—meaning "no bottom"—this zone does in fact contain over three quarters of the deep-ocean floor. The temperature here is not much lower than in the Bathypelagic Zone but the pressure increases to around 11,000 pounds per square inch. Very little life can be found at this depth, most of the creatures that have managed to adapt to the overwhelming pressure are invertebrates such as the deep-water squid and octopus.

The *Hadalpelagic Zone* (also known as "the trenches") is the deepest layer of the ocean, extending down from 19,700 ft (6,000 m) to the very bottom of the sea. The deepest point in the world's oceans is the Mariana Trench, off the south coast of Japan. Here the ocean reaches down to a depth of around 36,000 ft (10,970 m), this is further below sea level than Mount Everest is above it. The temperature is only just above freezing and the pressure exerted by the water above is an extraordinary 16,000 pounds per square inch. Even under these extreme conditions some life, such as tubeworms and jellyfish, exists.

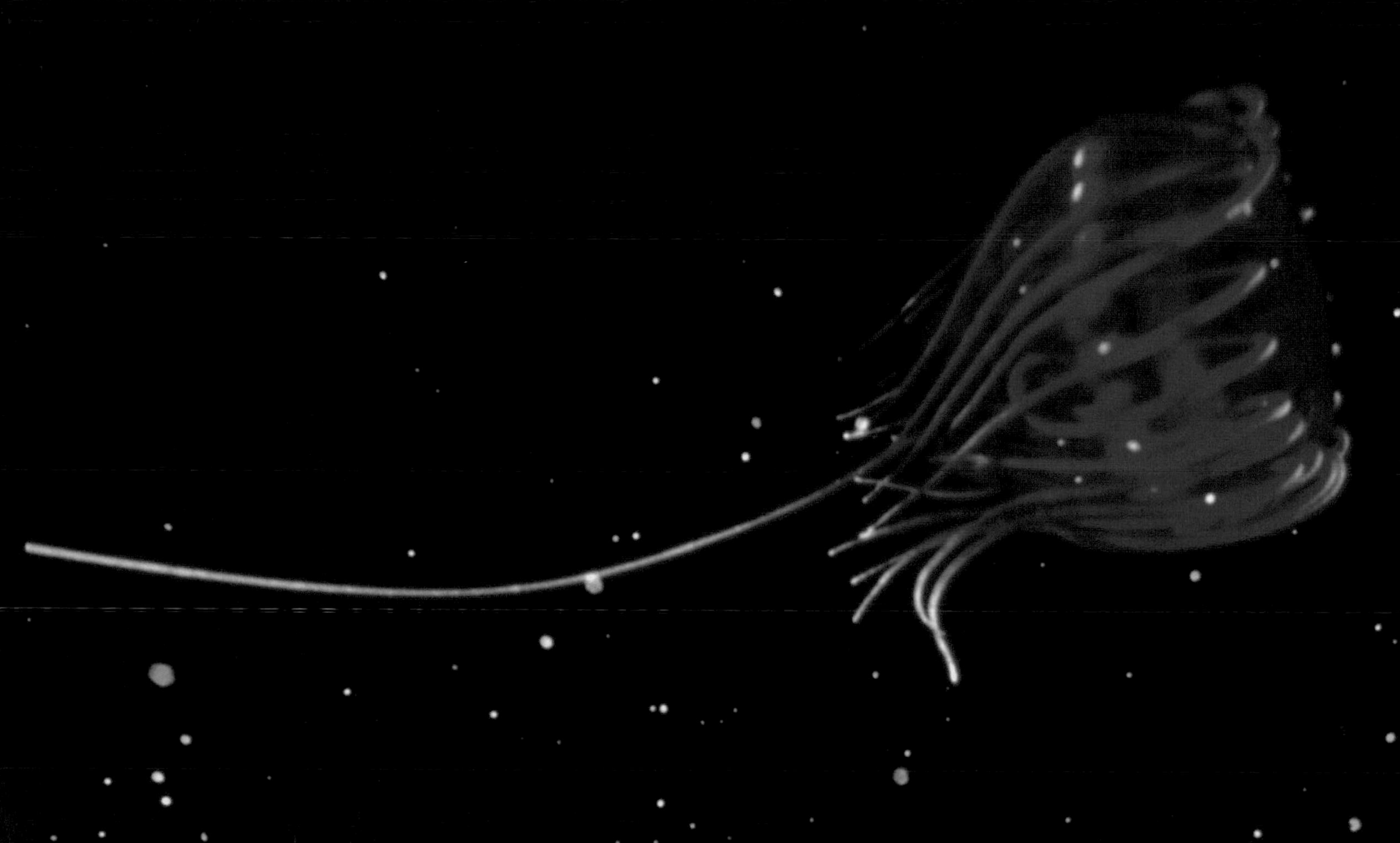

The Water Cycle

The significance of the oceans to life on the planet cannot be overstated; the oceans are where life on Earth originated and without water life as we know it could not continue to be sustained.

The water cycle (also known as the "hydrologic cycle") is the continuous process by which water circulates from the Earth's surface, up into the sky, and back down again. The main driving force of this process is the sun—heat from the sun causes water to evaporate (turn to from liquid to vapor) and pass into the atmosphere. This accounts for 90% of the water vapor that passes into the atmosphere from the surface of the planet, but there are two other ways in which this occurs—the majority of the remaining 10% is provided by the evaporation of water from plant leaves (known as transpiration) with the balance coming from ice and snow turning directly into water vapor with no liquid stage in-between (in a process called sublimation).

As the water vapor passes up into the air it is cooled and condensation occurs (i.e. the vapor begins to turn back into liquid droplets). These tiny droplets then collect on minuscule dirt particles of dust and dirt in the atmosphere and form clouds. When the droplets that make up the clouds grow large enough they fall back down to earth as precipitation (that is to say water that is released from clouds in the form of rain, hail, sleet, or snow) and the cycle begins again.

The Ocean Food Chain

The oceans contain some of the greatest diversity of life forms on the planet, from microscopic organisms such as zooplankton to the largest known animal to have lived, the blue whale.

The ocean food chain begins with phytoplankton—single-celled plants that float on the ocean surface—that convert energy from sunlight into oxygen and carbohydrates (this is called photosynthesis). Zooplankton—tiny floating animals such as krill and the larval stages of some crustaceans—drift through the sea grazing on the phytoplankton and are in turn fed upon by small fish. These small fish are eaten by bigger fish and at the top of the food chain are predatory fish, such as sharks and mammals such as seals. Although in general it is true to say that larger fish and animals prey on smaller creatures one or two steps down in the food chain some of the very largest sea creatures, including the blue whale feed almost exclusively on zooplankton.

As well as being the basis of the ocean food chain phytoplankton also has an equally important role to play in the preservation of life on land. Almost half of the oxygen that we breathe is generated by the photosynthesis carried out by phytoplankton.

Coral Reefs

Although coral reefs occupy only a small fraction of the Earth's oceans they are the most biologically diverse eco-systems on the planet; on land only the tropical rainforests can come close to rivaling them for bio-diversity.

Coral reefs generally develop in the shallow, warm waters of tropical seas. Usually near land, most are to be found between latitudes 30° north and 30° south. The largest coral reef in the world—so large that it is clearly visible from space—is the Great Barrier Reef, off the northeast coast of Australia. The Great Barrier Reef is a reef system made up of nearly 3,000 separate reefs, is over 1,250 miles (2,000 km) long and covers an area of roughly 135,000 square miles (350,000 square km).

Coral reefs can take hundreds, even thousands, of years to grow. Tiny animals called polyps that have a hard, calcium exoskeleton form them. As generation upon generation of polyps live and die their skeletons pile up on top of each other and along with the skeletal remnants of other reef-dwelling creatures these remains gradually build up to create large reef systems.

These reef-building corals are known as hard corals and include brain, elkhorn, and staghorn coral. Soft corals—such as sea whips, and sea rods—have no exoskeleton and thus do not contribute the growth of the reef, they do however come in a huge variety of dazzling colors, adding greatly to the beauty of the reefs.

Left: Situated southeast of the Phillippines, the rock islands of Palau, rise from the turquoise waters of the Pacific Ocean and are home to pristine coral reefs, dense jungles, and ancient caves.

Below: Galapogos fur seals (*Arctocephalus galapgoensis*) are the smallest of all fur seals and are seen only in the Galapagos islands. They are a protected species due to extensive hunting during the 19th century and were classed as vulnerable by the 2000 IUCN Red List.

As well as the corals that form them, the reefs are home to an extraordinary variety of sea-life including sponges, nudibranchs (that look like colorful slugs), clown fish, parrot fish, crabs, lobsters, octopi, and sea snakes.

The Five Oceans of the World

The *Pacific Ocean* is the largest of the world's oceans. It covers over a quarter of the Earth's surface, an area in excess of 63 million square miles (163 million sq km), is over twice the size of the second largest ocean (the Atlantic) and covers more of the Earth's surface than all of the land put together. The Pacific Ocean also contains the deepest point in all of the world's seas — Challenger Deep in the Mariana Trench.

The Pacific Ocean contains over 30,000 islands, the largest of which is New Guinea, however, the total area covered by all of these islands accounts for less than one percent of the ocean's total surface area.

Endangered species that can be found in the Pacific include sea lions, sea otters, dugong, and sea turtles. Only in recent years has a ban on whaling eased the steady decline of the whale population. The *Atlantic Ocean* is the second largest of the world's oceans; it covers around one fifth of the Earth's surface, an area of over 30 million square miles (77.7 million sq km). The deepest point in the Atlantic is Milwaukee Deep in the Puerto Rico Trench; situated 100 miles (160 km) north west of Puerto Rico, it has a maximum depth of 27,493 ft (8,380 m).

The Mid-Atlantic Ridge is a submarine mountain range that bisects the sea floor of the Atlantic. Stretching for around 10,000 miles (16,000 km) it reaches a breadth of 1,000 miles (1,600 km) at its widest points. Some of the mountains in the Mid-Atlantic Ridge reach above sea level and form islands such as St. Helena and Tristan da Cunha, as well as island groups such as the Azores.

Endangered species that can be found in the Atlantic include the manatee, sea lions, and seals.

The *Indian Ocean* is the third largest of the world's oceans; it occupies roughly one fifth of the total area of the Earth covered by oceans, an area of just over 28 million square miles (72.5 million sq km). The deepest point in the Indian Ocean is the Sunda Deep of the Java Trench, off the southern coast of Java, which has a maximum depth of 24,442 feet (7,450 m).

One of the most devastating natural disasters of modern times occurred in 2004 when the shockwaves from the undersea Indian Ocean earthquake caused a series of tsunamis—with

Left: Atlantic spotted dolphins (*Stenella frontalis*) are found in the tropical and subtropical waters of the Atlantic Ocean. They are very vocal and active at the surface and will often group with other types of dolphins. A typical family group could consist of 50 animals though smaller groups of six to ten are more common.

Above right: Skunk anemonefish (*Amphiprion sandracinos*) are found in shallow lagoons and on reefs at depths of approximately 9 ft (3 m) to 65 ft (20 m). This pair were photographed in the Andaman Sea.

waves of up to 100 ft (30m) high—that killed over 200,000 people and devastated large coastal areas of Indonesia, Sri Lanka, Thailand, India, and the Maldives.

The *Southern Ocean* is the fourth largest of the world's oceans; it extends out from the coast of Antarctica north to 60° south latitude, covering an area of over 7.7 million square miles (20 million sq km) that totally encircles the continent. The deepest point in the Southern Ocean is in the South Sandwich Trench, located west of the Mid-Atlantic Ridge between South America and Antarctica, where the deepest point is 23,737 ft (7,235 m).

The IHO took the decision the demarcate the Southern Ocean because it believed that the area it encompasses contains a sufficiently different eco system to the three oceans that it was formerly part of (the Indian, Atlantic, and Pacific) to warrant the creation of a new ocean to distinguish it from them.

The *Arctic Ocean* is the both the smallest and the shallowest of the world's oceans; covering an area of just under 5.5 million square miles (14.25 sq km) it is only just over a quarter the size of the next smallest, the Southern Ocean. The average depth of the Arctic Ocean is only 3,240 ft (987 m).

The central surface of the Arctic Ocean is covered by a constantly moving icepack that doubles in size in the coldest winter months. Recently scientists have become concerned that the summer ice covering is shrinking on a yearly basis, leading to speculation that it could disappear completely by the year 2060.

Top: Walrus (*Odobenus rosmarus*) Bering Sea, Alaska. Renowned for their tusks, which can grow up to 3 ft (1 m) in length, their skin can be over an inch (4 cm) thick and the blubber beneath can reach 6 inches (15 cm) in thickness. They inhabit ice floes in the Arctic and move south in the winter as the ice expands.

Above: Pacific harbor seals (*Phoca vitulina richardsi Linnaeus*) are gregarious animals and when not actively feeding they will rest on ice floes as seen here. They do not tend to venture more than 12 miles (20 km) offshore and are found north of the equator in both the Atlantic and Pacific oceans.

Left: Humpback whale (*Megaptera novaeangliae*) mother and calf in Hawaii, North Pacific. Breeding takes place in tropical waters during the winter months and typically each female will bear a calf once every two to three years. The gestation period is one year and a calf will nurse for about a year.

Above: Spotted porcellanid crab (*Neopetrolistes maculatus*) photographed in Wakatobi National Park, Sulawesi. At 5,300 square miles (13,900 sq km) Wakatobi is the second largest marine protected area in Indonesia.

Left: Pufferfish (*Arothron sp.*) are also known as blowfish and can turn themselves into an inedible ball several times their normal size. Almost all pufferfish contain tetrodotoxin, a substance produced within their systems which tastes foul. While often lethal to fish to humans it is deadly: 1,200 times more poisonous than cyanide. Each pufferfish carries enough toxin to kill 30 adults and there is no known antidote. Found mainly in tropical and subtropical ocean waters they range in size from 1 inch (2.5 cm) to the freshwater giant puffer which can reach 2 ft (61 cm).

Right: Two spinecheek clownfish (*Premnas aculeatus*) resting among the tentacles of an anemone.

Above: The Australian giant cuttlefish (*Sepia apama*) is found in waters at depths of less than 3 ft (1 m) to around 320 ft (100 m) off southern and eastern Australia. It can grow to up to 3 ft (1 m) in length, weighs up to 6.5 pounds (3 kg), and changes color to match its background in order to avoid detection.

Top left: This juvenile leafy seadragon (*Phycodorus eques*) is two to three months old. A relative of the seahorse, seadragons use their leaf-like appendages as camouflage. They inhabit the clear, temperate waters of Southern Australia at depths of 16 ft (5 m) to 49 ft (15 m). Having become endangered due to habitat destruction and demand for aquarium specimens, they have now been officially protected by the Australian government. This specimen was photographed in Wool Bay, South Australia.

Bottom left: A southern blue ring octopus (*Hapalochlaena maculosa*) female with eggs. Weighing a mere 1 ounce (28 grams) the species can be found in the shallow coral and rock pools of Australia. Be warned the blue rings only appear when the animal feels threatened and, if bitten, the octopus injects a venom containing tetrodotoxin, which can be fatal.

Right: This brain coral (*Platygyra*), photographed on the Great Barrier Reef, Queensland, Australia, shows blue bioluminescence under fluorescent light.

Next page: Pygmy sweepers (*Parapriacnathus ransonneti*) are also known as glass fish or golden sweeper and form huge groups. Swimming in formation they shoal together during the day and disperse to feed at night. They are they are each about 4 in (10 cm) long and are found in the Indian and western Pacific oceans.

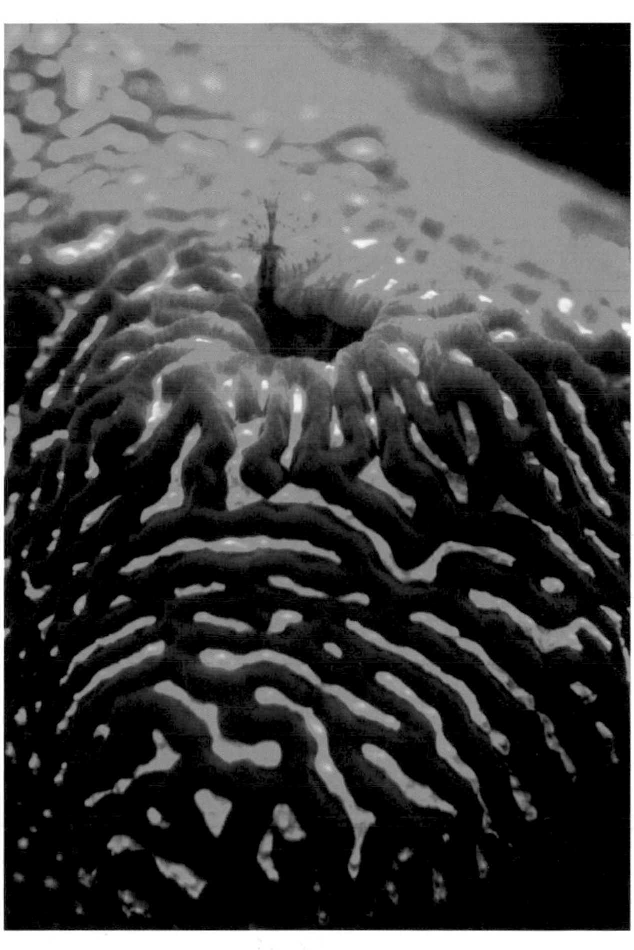

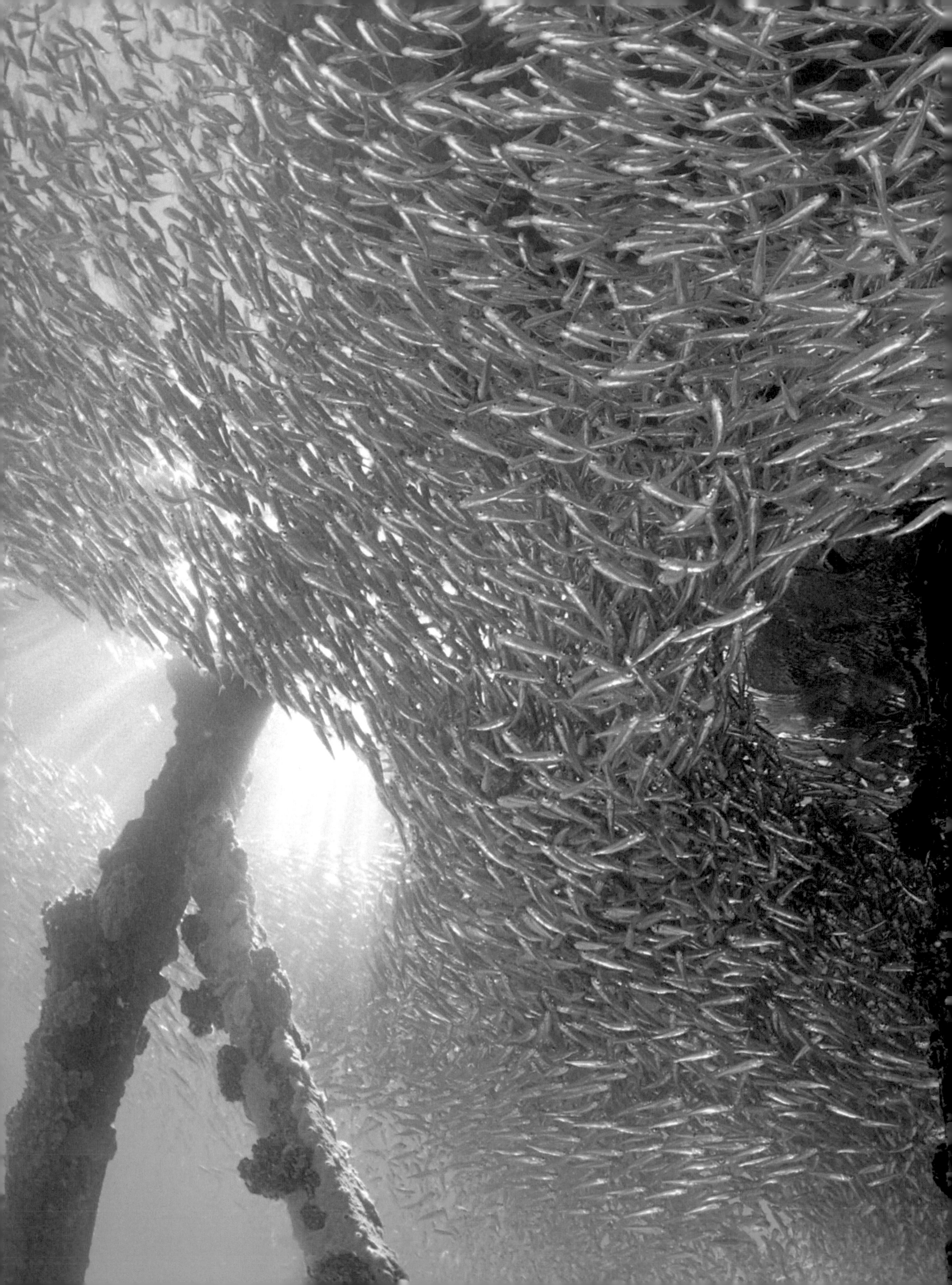

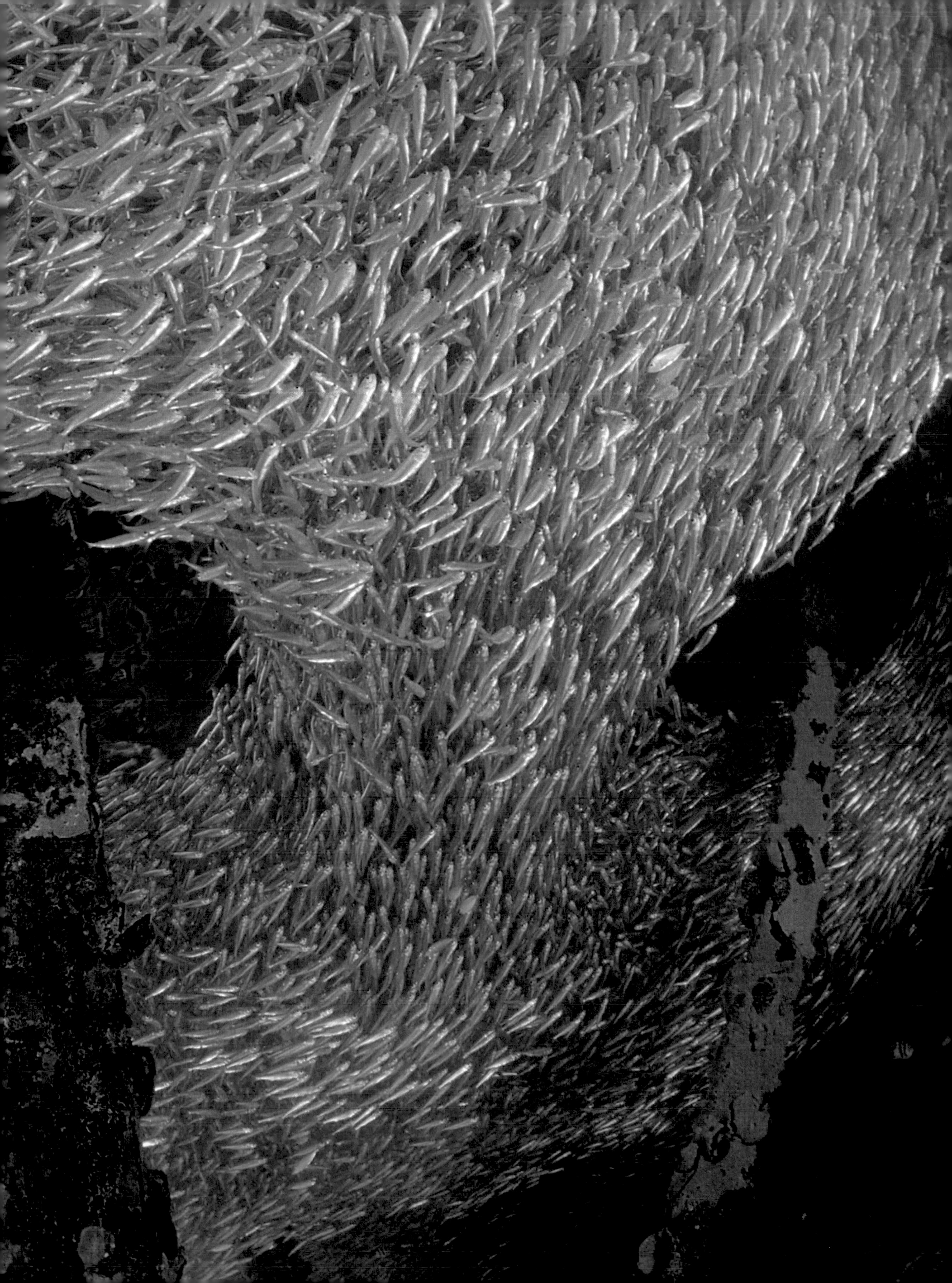

Main image: Hard corals are corals that contain a hard calcium skeleton due to the secretion of calcium that they produce. These examples are at Ras Mohammed in the Red Sea, Africa.

Below left: Crown jellyfish (*Netrostoma setouchina*) Ras Mohammed, Red Sea, Africa.

Below middle: A coral reef with colorful soft corals (*Alcyonacea, Dentronephtya sp.*). Soft corals occur in all reef habitats, but show impressive growth in deeper parts of the reef. They have a flexible skeleton and come in a dazzling array of colors and colony shapes. These examples were caught on camera at Ras Mohammed in the Red Sea, Africa.

Below right: Lionfish (*Pterois voltans*) are found predominantly in the Indian and Pacific oceans and have several venomous dorsal spines. Although painful, the sting is not fatal to humans.

Previous page: The green sea turtle (*Chelonia Mydas*) is one of the largest marine turtles, reaching between 2.2 ft (70cm) and 5 ft (1.5 m) in shell length and weighing up to 440 lbs (200 kg). Protected from exploitation in most countries they are recognized as endangered by the International Union for Conservation of Nature (IUCN) Red List. They inhabit the waters of the Atlantic, Pacific, and Indian oceans and this specimen was found in the Pacific Ocean, off Borneo.

Right: Whale sharks (*Rhincodon typus*) are found worldwide in tropical and warm temperate oceans. They are a filter feeder, feeding on phytoplankton, algae, plankton, krill, and small squid or vertebrates. The whale shark can grow up to 46 ft (14 m) in length and weigh up to 15 tons, while its mouth can be up to 4 ft (1.4 m) wide. This photograph was taken off Darwin Island, Galapagos, Pacific Ocean.

Far right: The blue whale (*Balaenoptera musculus*) is believed to be the largest animal to have ever existed. They can grow to a length of more than 98 ft (30 m) and weigh in excess of 100 to 150 tonnes. Found in all oceans worldwide, they feed mainly on krill and tend to live in pods of two or three individuals though have been seen in groups of 60 or more.

Above: The great white shark (*Carcharodon carcharias*) can reach lengths of more than 20ft (6 m) and weigh more than 4,500 lbs (2,000 kg). They swim at speeds of up to 43 mph (69 kph), have acute hearing and good eyesight, plus a keen sense of smell, and are the world's largest-known predatory fish. The great white lives in almost all coastal and offshore warm waters, with greater concentration off the coasts of Australia, South Africa, California, and Mexico's Isla Guadalupe. One of the densest known populations is around Dyer Island, South Africa, where this photo was taken.

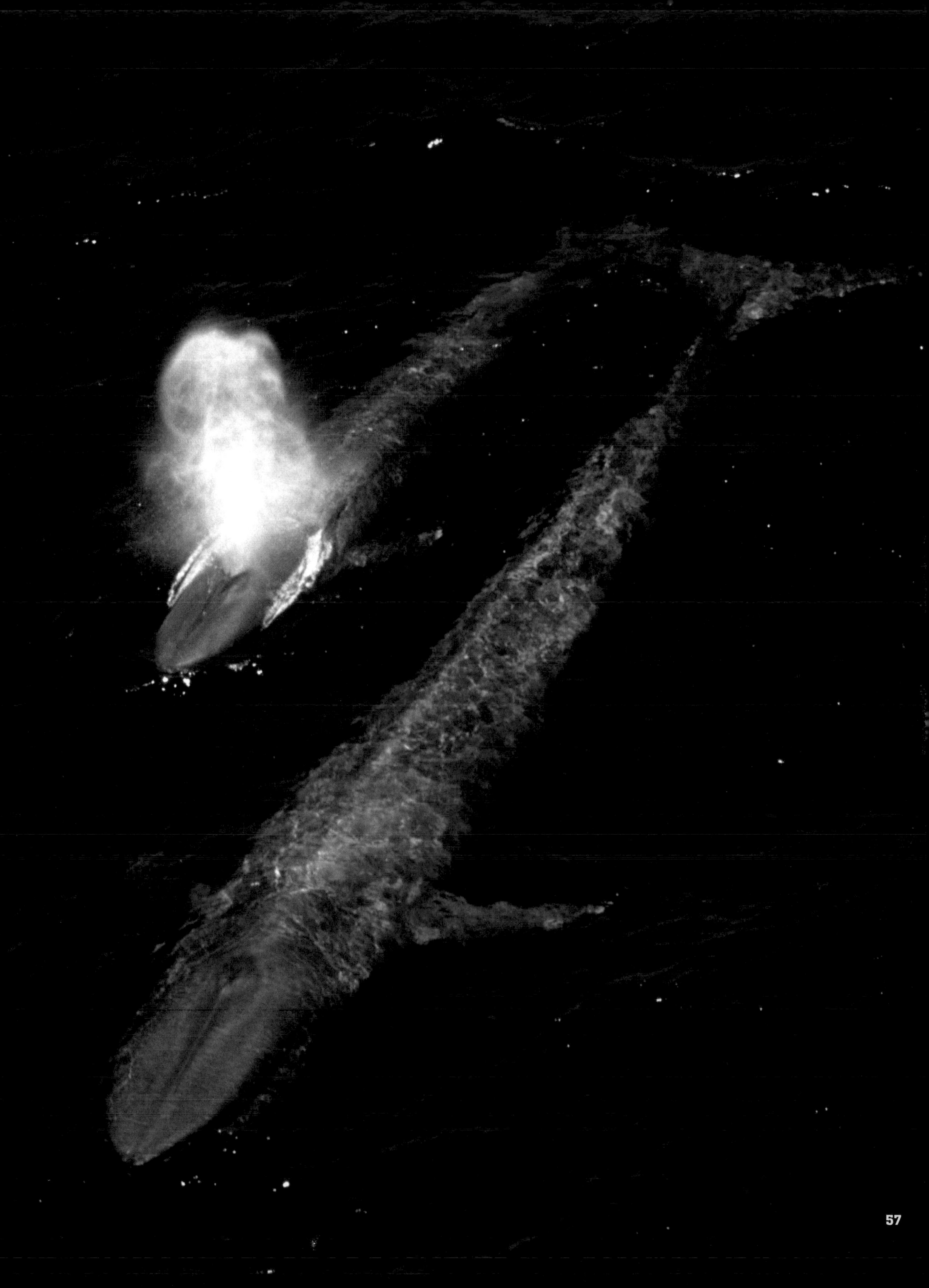

Deserts

Previous page: The sand dunes of Erg Chebbi, Merzouga, are formed by windswept sand with little or no vegetation and can reach heights of 490 ft (150 m).

Left: These sand dunes in Death Valley National Monument Park, California, are situated in the northernmost area of the Mojave Desert. Seen here in August, the desert's hottest month is July, when temperatures reach 120 °F (49 °C) and it is at its coldest in December, when temperatures fall as low as 0 °F (-18 °C).

Right: Spring in the Mojave Desert. The California poppy appears between February and May after the annual rainfall. This photograph shows the Antelope Valley Poppy Reserve, California.

Major deserts can be found on five of the world's seven continents. They cover just over a third of the total land surface of the planet—an area of just over 19 million square miles (49 million sq km). We tend to think of deserts as hot, dry, and barren places with little life—indeed the word "desert" comes from the Latin *desertum*, meaning "abandoned place"—and as a broad definition this does generally hold true. However, as with mountains there is no universally accepted definition of exactly what constitutes a desert. Most areas that are generally recognized as deserts receive less than 400 mm (16 in) of precipitation per annum, but another definition used by some experts is that a true desert receives less than 250 mm (10 in) per annum and those areas that receive between 250 mm (10 in) and 400 mm (16 in) are designated as semideserts.

North American Deserts

The *Great Basin Desert* is the largest desert in the United Sates (the other three being Chihuahuan, Sonoran, and Mojave deserts) and covers an area of 190,000 square miles (492,000 sq km). It is bordered by the Columbia Plateau in the north, the Wasatch Mountains in the east, the Mojave Desert in the south, and the Sierra Nevada range in the west. The majority of the Great Basin Desert is located within the state of Nevada—indeed the desert occupies most of the state—with most of the remaining portion occupying into the western half of Utah as well as small portions of California, Idaho, Oregon, and Wyoming.

The Great Basin Desert is a region of high valleys, basins, smaller mountain ranges, and dry lakebeds (playas). It is commonly referred to as a "cold desert" as, due to its more northerly latitude, the temperature here does not reach as high as in the other three North American deserts.

The valleys of the Great Basin receive little rainfall throughout the year; an average of 6 to 12 inches (150 to 300 mm) per annum. This is mainly due to the presence of the Sierra Nevada mountain range in the west, which forms a huge natural barrier blocking the moisture-bearing North Pacific winds from reaching inland. The majority of precipitation that falls does so in the winter in the form of snow.

One of the most notable features of the Great Basin is that it is an "endorheic" basin—that is to say that none of the precipitation that falls in it drains out to the sea; it either falls on land and evaporates or drains into rivers that flow into inland lakes that have no natural outlet to the sea (i.e. the Humboldt Lake in northwestern Nevada).

The *Mojave Desert* is the smallest of the four North American deserts and is largely situated in southern California, though it also reaches into areas of Nevada, Utah, and Arizona, covering a total area of some 25,000 square miles (65,000 sq km). The desert was named after the Mojave Native Americans, whose traditional tribal lands are located along the banks of the Colorado River in Arizona and California.

The Mojave is home to the hottest, lowest, and driest area in the North American continent—Death Valley. The lowest point in Death Valley (also the lowest point in the Western Hemisphere) lies in the Badwater Basin, 282 ft (86 m) below sea level. The highest temperature ever recorded in Death Valley is 135 °F (57 °C), on July 10, 1913.

The weather in the Mojave ranges from extreme highs in the summer to sub-freezing in the winter and the area has a noticeably seasonal climate. Summer is, naturally, the hottest and driest season in the Mojave and temperatures in the valleys can reach as high as 120 °F (49 °C). While the Mojave is the driest of the American deserts, thunderstorms drawn in from the Gulf of Mexico are not uncommon in the southwestern areas during the summer months.

The winter brings more frequent precipitation in the form of both rain and, occasionally, snow. In the higher regions the temperature has been known fall as low as 0 °F (-18 °C).

Autumn and spring are less extreme, with temperatures in the valleys usually ranging between 70 °F (21 °C) and 90 °F (32 °C) during autumn and occasionally reaching above 100 °F (38 °C) during the spring, the early months of which see an increase in rainfall that tails off toward the end of the season.

Along the western areas of the Mojave high winds of up to 50 mph (80 kph) are not uncommon and a number of wind farms have been established in the region to take advantage of this and provide ecologically friendly power.

While vegetation in the Mojave is sparse some species of cacti can be found here and, though trees are particularly rare, the desert is home to the Joshua Tree National Park that contains thriving forests of its namesake trees.

South American Deserts

The *Patagonian Desert* is the largest desert in South America. It covers an area of 260,000 square miles (673,000 sq km) the majority of which is located in the south of Argentina, though parts of the desert extend into Chile.

The Andes Mountains, to the west, are the main reason for the aridity of the area; the Patagonian is a classic example of a rain shadow desert. Just as the Sierra Nevada mountain range blocks moisture-bearing North Pacific winds from reaching the Great Basin Desert in the U.S. so the Andes stop the flow of such winds from the South Pacific reaching inland.

The Patagonian Desert has two distinct regions, the north and the south. The northern region is the wetter of the two, receiving annual rainfall of between 3.5 in (90 mm) and 17 in (430 mm). In the summer, temperatures reach up to 113º F (45° C) and in the winter they can fall to 12° F (-5° C). The climate in the southern region is harsher than that of the north. Here the temperature can drop to as low as -27° F (-33° C) in the winter and heavy snowfall is not uncommon at this time. In the summer the highest temperature reached it around 93° F (34° C), however, even in the summer it is not unknown for frosts to occur. Average annual precipitation (made up of rain and snow) ranges from 5 in (127 mm) to 8 in (203 mm).

The *Atacama Desert* in northern Chile is one of the driest places in the world. It is bordered on the west by the Cordillera de la Costa (a range of low coastal mountains) and to the east the foothills of the Andes. The boundaries of the Atacama are not precise but it stretches somewhere between 600 m (966 km) and 700 m (1,127 km) from the Chile/Peru border in the north to the Loa River in the south. There are areas of the Atacama that have received no rainfall at all since records began.

The extreme aridity of the Atacama is in part due to the rain shadow cast by the two mountain ranges to its east and west and also the effects of the Humboldt Current—a sea current that carries cold water up from the Antarctic to the seas off the coast of Chile. This colder water at the sea surface causes warmer air to rise, forming fog and clouds but little, if any, rain. These factors also combine to make the temperatures in the Atacama low, with average summer temperatures in reaching around 65 °F (18 °C).

Left: A cottonwood tree (*Populus deltoides*), also known as the "water tree," photographed in the Chihuahuan Desert, White Sands National Monument Park, New Mexico. In incredibly harsh conditions the Cottonwood has managed to survive by taking root in the sparse interdune flats and accesses water by extending roots into the shallow water table below.

Above right: Welwitschia (*Welwitschia mirabilis*) in the Namib Desert. An ancient plant with long leaves that gradually become torn and shredded by the constant drying desert winds, welwitschia survives the dry conditions by absorbing water left on its leaves from morning dew and fog that flows over the Namib Desert.

African Deserts

The *Sahara Desert* is the largest in the world, covering an area of 3.5 million square miles (9 million sq km) in the north of Africa. This vast desert stretches from one side of the African continent to the other, covering large areas of Algeria, Chad, Egypt, Libya, Mali, Mauritania, Morocco, Niger, Western Sahara, Sudan and Tunisia. It is bordered to the north by the Atlas Mountains and the Mediterranean Sea, to the east by the Red Sea, while to the south it stretches down to latitude 16° N.

Flat sands and the more impressive sand dunes that are commonly associated with the desert environment cover only a quarter of the Sahara. It also contains many mountains and mountain ranges —over a third of the desert is mountainous—with the highest point being the 11,204 ft (3,415 m) high Mount Koussi in the Tibesti Mountains in Chad.

The Sahara is one of the harshest environments in the world; precipitation is sporadic and between three quarters of an inch (20 mm) and 4 in (100 mm) per annum falls in most regions, much of it during thunderstorms. In the sandy areas, dust devils and sand storms are frequent and are a deadly hazard. The highest temperature ever recorded on the planet was in the Sahara: 136° F (58° C) at Al-Azizia, in Libya on September 13, 1922.

The *Namib Desert* is generally considered to be one of the oldest in the world; it is thought to have existed for over 55 million years. It runs for 1,200 miles (1,900 km) along the west coast of Africa. The far northern and southern sections of the Namib are located in Angola and South Africa respectively while the greater part of the desert occupies the entirety of the Namibian coastline.

The name "Namib" comes from the Nama language and means roughly "place where there is nothing." This is certainly apt as the Namib is almost entirely uninhabited apart from a few isolated coastal towns such as Walvis Bay and Swakopmund.

The Namib is an extremely arid desert. Average annual precipitation is as low as half an inch (13 mm) in coastal areas and even inland this only rises to 2 inches (51 mm) per annum. As in the Sahara, rainfall is sporadic and it is not unknown for whole years to pass without any rain, and that which does fall is, again, frequently in the form of thunderstorms.

Because of this almost total lack of rain much fauna in the coastal regions of the Namib is reliant on the thick fog that blows in from the Atlantic, covering the dunes with much-needed moisture. One of the more remarkable plants to be found in the Namib is *Welwitschia mirabilis*, thought to be the longest-living member of the plant kingdom—some of the oldest are believed to have lived for as long as 2,500 years. The Welwitschia is unique to the area, stands up to 5 ft (1.5 m) tall and has just two fibrous leaves which continually grow and endure for the lifespan of the plant.

Australian Deserts

The *Great Victoria Desert* is the largest of the Australian deserts, covering an area of around 164,000 square miles (424,000 sq km) in Western and South Australia.

Precipitation, always in the form of rain, is sporadic—the average annual rainfall varies from 8 to 10 inches (200 to 250 mm)—and frequently falls during violent thunderstorms. However, when it does rain parts of the desert are transformed, with fields of wildflowers springing up almost overnight.

Daytime temperatures in the summer reach up to 104 °F (40 ºC) and in the winter it is still as high as 73 ºF (23 ºC), but at night in the winter it can fall to freezing level and frosts are not uncommon.

The *Simpson Desert* covers an area of around 55,000 square miles (143,000 sq km) in central Australia. The majority of the desert is in Northern Territory but parts of it also extend into Queensland and South Australia; it is bordered to the north by the MacDonnell Ranges and the Plenty River, to the east by the Mulligan and Diamantina rivers, to the south by Lake Eyre and to the west by the Finke River.

One of the most famous features of the Simpson Desert is its dunes; it contains some of the longest parallel sand dunes in the world, which are held in place by the desert scrub and other vegetation that grows there. The largest is "Big Red," which stands almost 130 ft (40m) high.

The Simpson Desert is home to a number of rare species including the water-holding frog—a seldom seen creature which can spend several years in its burrow, living off water stored in its bladder, before emerging to breed following heavy rainfall—and the fat-tailed marsupial mouse.

Because of its ecological importance large areas of the Simpson have been given protected status and it is home to three large reserves; Queensland's Simpson Desert National Park, South Australia's Simpson Desert Conservation Park, and the Simpson Desert Regional Reserve.

Above: Wave-like dunes of gypsum sand in the Chihuahuan Desert, White Sands National Monument, New Mexico. Gypsum is highly soluble in water so is rarely seen in this form, but in an environment where there is no natural water outlet to the sea when the sand absorbs rain water it will pool into a lake that evaporates to create flats, which will then gradually break back down into gypsum sand.

Left: Salt encrusted outcrops in the Atacama Desert, South America. Even though this is a cold desert, with temperatures between 32° F (0° C) and 77° F (25° C), evaporating lakes leave behind mineral salts that create the deposits that form the structures pictured here.

Asian Deserts

The *Gobi Desert*, which takes its name from the Mongolian word *gobi* (meaning "waterless place"), covers large portions of both Mongolia and China and is the largest desert in Asia. The Gobi is bordered on all four sides by mountain ranges; to the north the Altai and Hangayn mountains, to the east the western extremes of the Greater Khingan Range, to the south the Pei and Yin Mountains, and to the west the eastern reaches of the Tien Shan Mountains. The desert is 1,000 miles (1,609 km) long and reaches widths of up to 600 miles (966km). In total, it encompasses an area of around 500,000 square miles (1,300,000 sq km).

Only the southern region of the Gobi, the Ala Shan Desert, is particularly sandy, much of the rest of the desert is made up of barren, rocky plains, low hills, dry riverbeds, and salt marshes.

The climate is one of great extremes; summer highs of 113° F (45° C) contrast with winter lows of 40° F (-40° C) and fluctuations in temperaturein a single day can be significant. Annual precipitation ranges from below 3 in (76 mm) in the west up to 8 in (203 mm) in the northeast.

The Gobi has been the site of a number of significant archaeological discoveries; in the 1920s a series of expeditions sponsored by the American Museum of Natural History discovered the fossilized eggs of a previously unknown species of dinosaur—the oviraptor—and in 1998 a joint expedition between the American Museum of Natural History and the Mongolian Academy of Sciences discovered the remains of an early relative of marsupials, the deltatheridium.

Previous page: An oryx (*Oryx beisa*) against the red-orange sands of the Namib Desert. Found primarily in the steppes and semi-deserts in eastern Africa, the oryx has developed the ability to raise its body temperature in extreme heat, preventing water loss through perspiration and thereby conserving water.

Above: Sossusvlei in the Namibia Desert is an old clay pan that has not had regular water since the river changed its course. Only the skeletons of old camelthorn acacias exist here, some of them well over 500 years old. It is enclosed on three sides by redish-orange sand mountains, some of which rise as high as 984 ft (300 m).

Above: The Gobi desert covers a vast area encompassing southern Mongolia and parts of China. Its land mass is estimated at 500,000 square miles (1,300,000 sq km) making it one of the world's greatest deserts. Unlike the Sahara's rolling dunes, its structure is typified by gravel plains and rocky outcrops. Its climate is extreme with temperatures ranging from 113°F (45° C) in the summer months to 40° F (-40° C) in winter.

Top right: The Kalahari Desert, Namibia, is characterized by four types of surface; Kopjes (which consist of exposed bedrock), sand sheets, longitudinal dunes, and pans. Encompasing an area of approximately 360,000 square miles (930,000 sq km) it covers Botswana, eastern Namibia, and the northern tip of South Africa.

Right: The meerkat (*Suricata suricatta*) is a member of the mongoose family and is found in southwestern Africa. In this photo three sentries stand on hind legs on the lookout for predators while the pack forages for food. If a predator is sighted the guards will emit a high-pitched call and the pack will scatter.

Main image: Tenacious annual desert herb (*Aseracaeae*) briefly colonizes a sand dune in the Simpson Desert before setting seed and dying in the summer heat to form wind-blown "rolly-polies." Otherwise inhospitable desert dunes can act as water storage if their core consists of finer moisture retentive material. For example, the perennial sandhill oak (*Grevillea stenobotrya*), shrubs tap deep moisture. Annuals survive the worst of the heat by means of their drought resistant seed stock and ability to germinate quickly.

Far left below: The thorny devil (*Moloch horridus*) is found in the deserts of central Australia. Not as fearsome as it looks, the lizard feeds only on ants and can eat thousands each day and it can also change color to camouflage itself against predators. The lizard collects water from dew in the grass between its scales. The fluid then runs down the lizard's body and into its mouth to drink.

Below middle: These bright orange streaks are recent fire scars. Beneath them can be seen the pattern of the sand dunes, the green tinge showing the layer of desert scrub that keeps these once shifting dunes static.

Below: The stumpy tail or shingle-back lizard (*Tiliqua rugosa*), native to New South Wales, is a slow-moving creature. It will bask in sunny areas in the early morning before foraging for food in the afternoon. For defence it relies on its armor and a bluff display, opening its mouth to reveal a blue tongue—that contrasts vividly with a pink mouth—to warn off predators.

Main image: One of the hottest places on Earth, the Danakil Desert in northern Ethiopia also features Lake Assal—Africa's lowest point at 515 ft (157 m) below sea level—which is the largest saline lake in the region. As there are no rivers flowing out of the lake, the water is ten times saltier than the ocean and is the most saline body of water in the world.

Above: The Atacama Desert, often described as the driest on Earth, is situated between the Andes mountains and the Pacific Ocean. There are areas where no rainfall has ever been recorded and there is no evidence of water running through for at least 100,000 years.

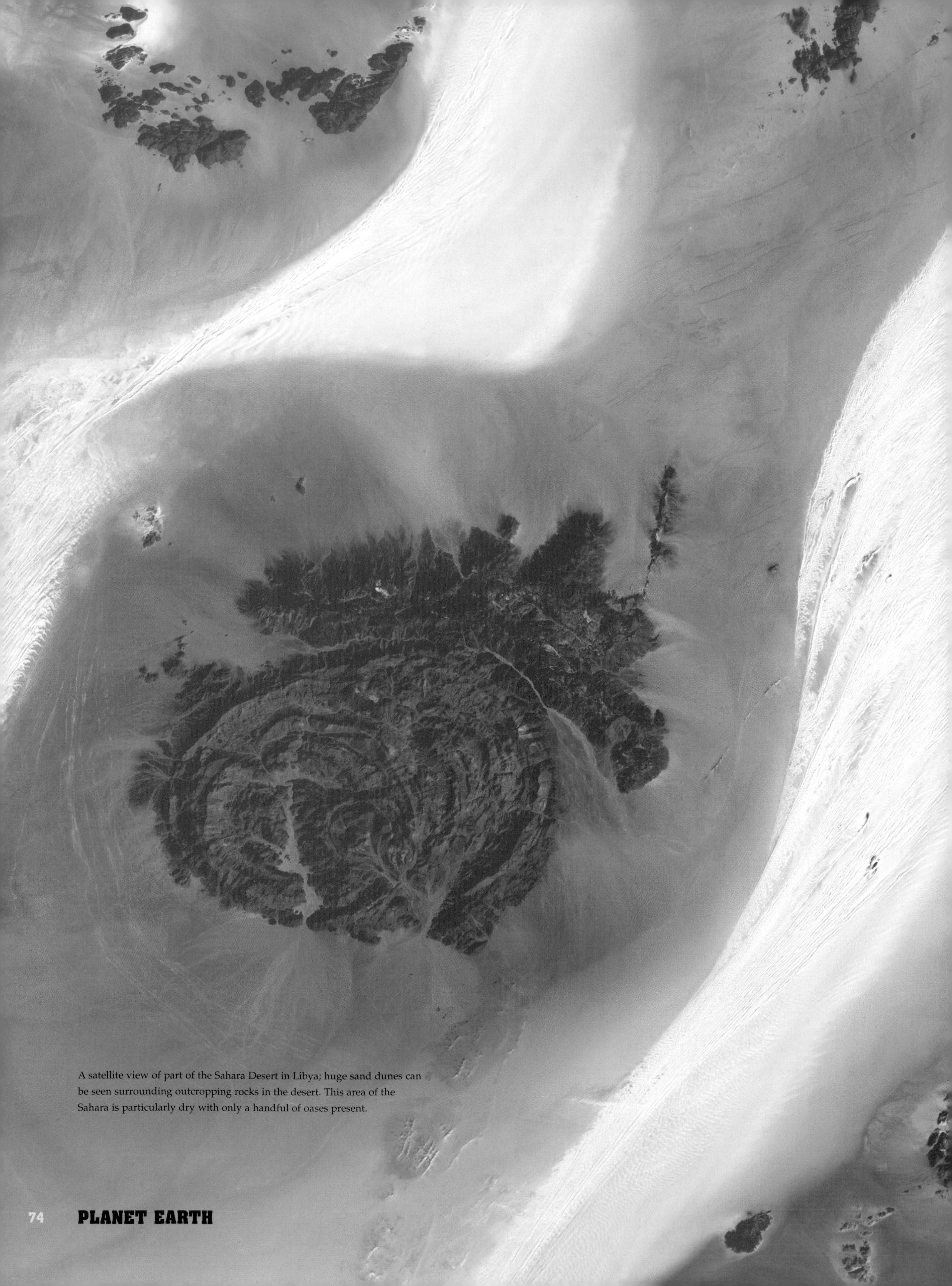

A satellite view of part of the Sahara Desert in Libya; huge sand dunes can be seen surrounding outcropping rocks in the desert. This area of the Sahara is particularly dry with only a handful of oases present.

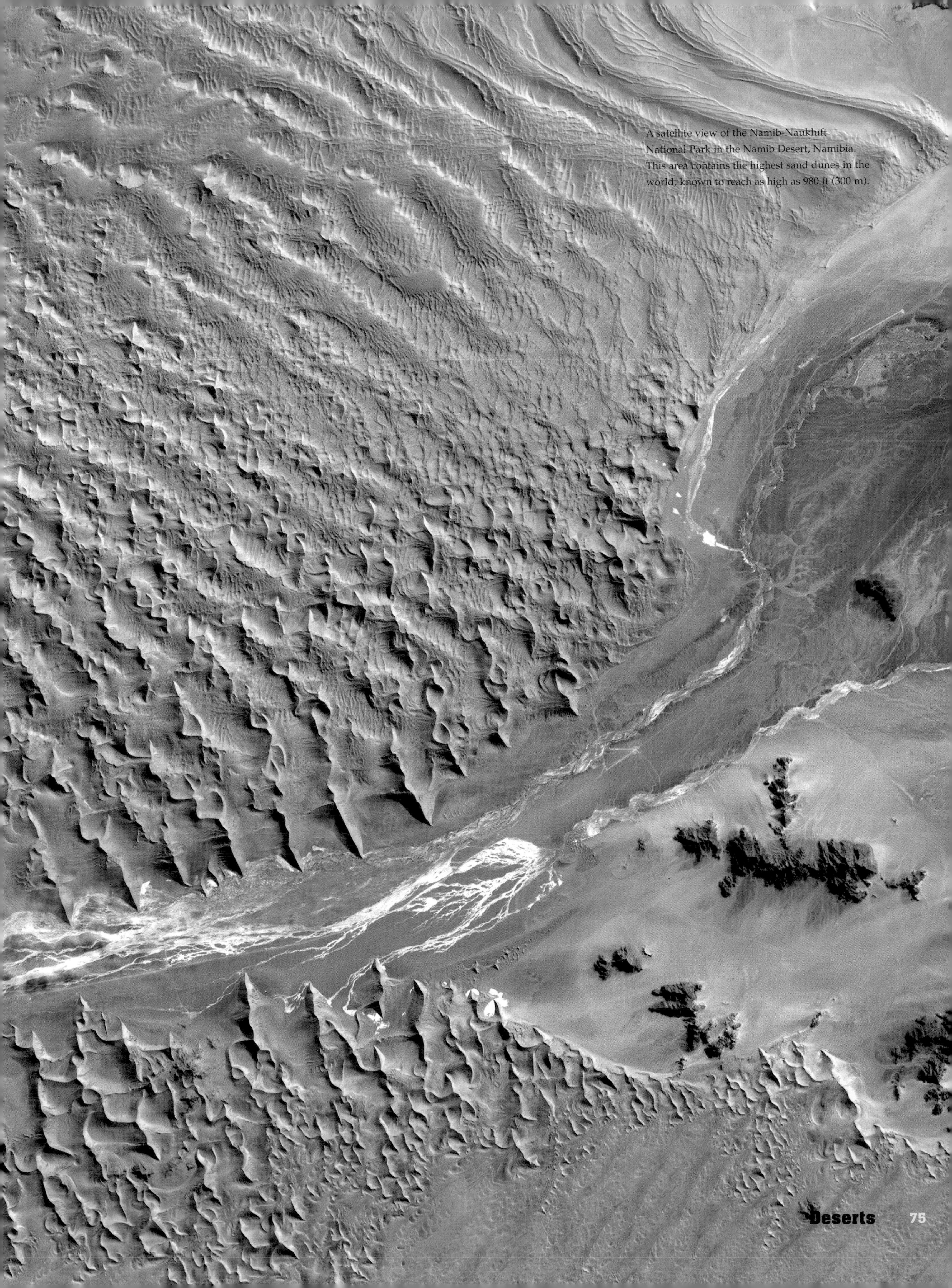

A satellite view of the Namib-Naukluft National Park in the Namib Desert, Namibia. This area contains the highest sand dunes in the world, known to reach as high as 980 ft (300 m).

Woodlands

Forests are defined as areas where high trees in dense proximity predominate the environment. They can be indigenous or planted by man—known as "anthropogenic." Where the tree canopy is more open and the trees are spaced further apart so sunlight can penetrate, the environment is defined as woodland. Ancient forests contain only tree species native to the area, while anthropogenic forests often contain non-indigenous species, for example the many conifers planted specifically for telegraph poles, railroad use, and the construction industry.

The forests themselves differ according to climate, altitude, and environment. Broadly speaking, forests and woodlands grow in environmentally friendly lands that also comfortably accommodate people—they are not found in extreme environments such as persistent cold regions like the Arctic and Antarctic, nor in predominantly dry and hot zones like deserts. Look at any range of mountains with tree-covered sides: at the same elevation suddenly there are no trees—this is known as the "tree line" where conditions no longer support growth. The transition is more gradual at the southerly extent of forests as the trees get shorter (due to less water) and more sparse until grasses predominate. But wherever they are found around the world, forests tend to be managed (unlike the denser and wilder rainforests).

Long ago the majority of the inhabitable lands across the Earth were tree-covered and our ancestors roamed under the leaf canopies in their search for food and shelter. Even as recently as eight thousand years ago ancient forest still covered over half the land. But as human populations grew the forests were chopped down to make way for settlements and early slash and burn agriculture. As the science and understanding of agriculture progressed, increasing areas of forest were cut down. In pre-industrial times forests were extensively logged for buildings and ship construction. In more recent times logging has been a major form of attrition—and while spawning a tree-planting regime in many areas of the world, logging has reduced still further the surviving pockets of ancient woodland and forest. Currently forests are estimated to cover only about 30 percent of the land surface of the Earth and without constant vigilance and legal protection many of these are in danger of disappearing. An estimated 20 percent of the world's ancient forests have disappeared in the last fifty years, greatly reducing biodiversity and endangering many species.

The decline of modern forests can be due to a number of factors including forest fires—lightning strikes or deliberate—wind damage, legal and illegal logging, and the corrosive effects of acid rain caused by pollution often hundreds of miles away. Additionally pest and disease damage or fungal infections can be devastating in isolated populations and in unusual circumstances insect populations can build up to such a level that they strip the trees of every scrap of foliage.

However, deforestation on a grand scale is almost entirely down to mankind. Following the Industrial Revolution in the mid-1800s the demand for timber exceeded supply and it is the environment that has paid the price. In the United States about 15 million acres (60,000 sq km) of woodland was removed by 1850 and by 1910 this figure had increased to a staggering 163 million acres (660,000 sq km), and the logging still continues, though not at such a rapid rate.

These days a potential greater risk for the survival of forests is climate change. This puts all types of forest at risk: as the subarctic zones get warmer and winters milder many pines will suffer distress and die. The species in the temperate zones will alter and perhaps progress north to follow the cooler weather they prefer and the trees in the Mediterranean zones will die out as their water supply declines and vanishes. Trees absorb huge amounts of carbon dioxide, one of the primary atmospheric pollutants, but if the trees disappear, more carbon dioxide will remain in the atmosphere and the planet will warm up even more.

Taiga

The "taiga" or boreal forests occupy the subarctic zone across North America, Northern Europe, and Asia and comprise the world's largest terrestrial biome. Most cold-tolerant forests are made up of coniferous trees (long-lived evergreen, cone-bearing trees), all of which with the exception of the larch (*Larix species*) are evergreen, and which can cope with the thin, poor soils and freezing winters of such high latitudes. Here the snow cover in winter can last for many months, and only similarly cold weather adapted plants and creatures exist. Relatively few species can thrive in such extreme conditions and it is here in the taiga that trees grow up to the tree line.

In less harsh taiga regions, where the winters are milder, a few small-leaved deciduous species such as willow, aspen, birch, and alder grow among the conifers. While at the warmest littorals of the taiga species such as maple, oak, and elm appear. Such forests still thrive across a vast area that includes

Previous page: A beech avenue in early November, Windsor Great Park, Berkshire, England. One of the most glorious trees of the European fall is the deciduous common beech (*Fagus sylvatica*), predominantly found on chalky and limestone soils across the cool temperate lands of Britain and most of Europe with the exception of Spain, Russia, Sweden, and Norway. Beech becomes fully mature at around 120 years by which time it will have achieved a maximum height of 131 ft (40 m); small yellowish flowers appear between late April and early May and develop into triangular winged nuts known as beech "mast" by fall.

Left: Heavily snow covered trees in the Bavarian Forest National Park. Together with its eastern neighbor the Bohemian Forest of the Czech Republic (Sumava National Park) this woodland is the largest area of contiguous protected forest in Europe. Special adventure trails wind through ancient fir trees, beech, and pines that soar over 164 ft (50 m) high and comprise the last vestiges of pristine mixed woodlands and mountain forests in central Europe.

Below: One of the most alluring animals which inhabit broadleaf and mixed forests is the highly endangered giant panda (*Ailuropoda melanoleuca*). Native to southwestern China each animal subsists on a frugal diet of 26 - 83 lb (12 - 38 kg) of bamboo shoots a day.

Alaska, Canada, the more northerly U.S. states of Maine, New Hampshire, upstate New York, Michigan and northern Minnesota, Russia, Sweden, Finland, parts of Norway, northern Kazakhstan and Hokkaido island in Japan. In the northern hemisphere the tree species of the subarctic zone are predominantly limited to members of the conifer family and include pines, spruces, larches, silver firs, Douglas firs, and hemlocks. In the southern hemisphere—particularly South America, New Zealand, Australia, and New Caledonia, the main species are Araucariaceae (such as the Monkey Puzzle Tree) and Podocarpaceae.

Temperate Broadleaf Forests

Warmer and wetter temperate lands contain deciduous forests made up of many and varied species of broadleaf trees which for around five months of the year are leafless allowing for a rich understory of smaller plants, insects, and animals. Many of these smaller plants follow a predictable seasonal calendar of growth, blossom, and seed or fruit set, before the leaves open on the trees above and the forests and woodlands return to dappled shade and even deep shadow.

Many different species of tree grow in such forests, in particular oak, birch, ash, maple, beech, sycamore, basswood, and hickory as well as many others, though a characteristic they share is that they are all slow growing (in contrast to tropical trees). These all grow fresh leaves in spring, flower to attract pollinators, and then develop seed or fruit which is then wind-scattered or distributed by animals and insects. In fall, the leaves drop to develop a rich layer of leaf litter on the ground and by winter the trees are bare, allowing plenty of light to reach the dirt. Such forests sustain a huge variety of co-dependant insects, birds, reptiles, and mammals all of which rely on the health and protection of the trees.

Mixed Evergreen and Deciduous Rainforest

Nearer to the equator, in the temperate latitudes where the mean temperature is much drier and hotter and the winters mild, the Mediterranean forest or temperate rainforest thrives. Another requirement is high rainfall, typically in the order of over 47 inches (1,200 mm) annually. These forests contain numerous broadleaf tree species; typically

Left: Cork Oak forest near Alpotal, Algarve, Portugal. The species—*Quercus suber*—is indigenous to southwest Europe and northwest Africa where it is highly valued for the properties of its outer bark. The cork is carefully harvested by hand every nine to twelve years in late spring or early summer; each tree can be productive for approximately 200 years, but it is only on the third harvest that the cork can be used for bottle stoppers.

Above right: The narrow-leaved Bottle Tree (*Brachychiton rupestris*) is a native of Queensland, Australia, where it grows up to 40 ft (18-20 m) in height. It gets its name from the bulbous appearance of its swollen base caused by storing gallons of water. This characteristic feature, however, does not develop until the tree is about fifteen years old.

many of these are evergreen with tough shiny leaves that retain their moisture despite the high temperatures. Often these trees grow to an impressive size which in turn has made them a considerable target for logging: at one time before the growth of human populations temperate rainforests covered vast areas of land, now they have been reduced by over 50 percent.

North America

Current estimates put the forested regions of the United States at 750 million acres (some six million square kilometers) but around 1.5 million acres (6,000 sq km) is being lost annually to urban development, especially in the South. The situation is better in the northern states where forests enjoy better state protection and are even being extended in some places.

Before the West was opened up, great swathes of Washington, Oregon, Northern California, and British Columbia were thickly forested with huge and impressive specimens of hemlock, spruce, Douglas fir, Sitka spruce, and Western red cedar. But these were rapidly depleted and taken not just for their timber but for the land on which they stood. Now less than ten percent of these original woods (or "old growth forest" as it is properly called) remain and even these are at risk from logging and need powerful protection from both local and national government to ensure their survival for future generations to enjoy. Such forests are ecologically important as they contain rare and threatened species, not just of trees and plants, but also of animals, birds, reptiles, and insects. When logged the forests become fractured and no longer provide essential ecosystems for wild animals and protected travel corridors so they can safely move from one area to another and increase their own genetic makeup.

The North American forests lie predominantly along the West Coast: they start in northern California, cover the numerous islands such as Vancouver Island, the Queen Charlotte Islands, and hundreds of smaller islands all the way up as far as the Tongass National Forest in southeast Alaska, where the Sitka spruce is the dominant species. This latter is an area of wilderness of some 17 million acres (69,000 sq km) of coastal wetlands and forest. Further south on the Olympic Peninsula of western Washington is Olympic National Park and the spectacular Hoh Forest. There are also extensive temperate forests in Oregon and small areas in the Rocky Mountains in northwest Montana. Many of these forests are now protected from logging by national park status and the only tree felling is done for conservation and environmental reasons.

Some of the most ancient and most impressive trees are the famous redwoods of Sequoia National Park, California, many of which are over a thousand years old. These vast trees thrive in the specific climate created by the warm and moist breezes blowing off the Pacific Ocean. Here the temperature averages 60° F (15° C) all year and as a consequence the trees keep growing slowly and steadily.

The Canadian boreal forest covers over a million and a half square miles (almost 4 million sq km), over half of which is cared for by sustainable management and intended for the timber trade. Any harvested areas have, by law, to be replanted to sustain the environment.

South America

Much of South America is covered with rainforest, but small areas in southern Chile and Argentina contain temperate forests. In Chile the ancient forests grow southern beeches and the unique monkey-puzzle tree. Both species are threatened by logging, fire, and habitat destruction through drought and flood. The coniferous monkey puzzle tree (Araucaria araucana) lives in coastal and Andean mountain ranges in Chile and some parts of Argentina and can live for over 1,000 years in its natural habitat. It has become the national tree of Chile and a protected species; it is also a protected species in Argentina.

Northern Europe

This area still contains extensive forests which stretch from parts of Ireland in the west, through parts of England and Wales, across Scotland, then Scandinavia, the Baltic states, much of Germany and central Europe, Russia—and especially Siberia—all the way to Asia. However, only three percent (Greenpeace figures) of this is old growth forest, primarily boreal forest in Scandinavia and on the western flanks of the Urals in Russia. In central Europe the Bavarian Forest National Park (Germany) and its eastern neighbor the Bohemian Forest (Czech Republic) together provide the largest area of contiguous protected forest in Europe. In this invaluable environment wildlife thrives even though the public have ready access to large areas.

Left: New England. The North American fall is one of the great spectacles of the natural world as deciduous trees prepare for winter. Occurring from around the first week of September and lasting until late October, these forests of green turn into firework shades of yellows, golds, oranges, rusts, and reds.

Below left: Eastern North American deciduous hardwood forest of white oak (*Quercus alba*), red oak (*Quercus rubra*), black oak (*Quercus velutina*), bitternut hickory (*Carya cordiformis*), and shagbark hickory (*Carya ovata*). Thanks to human activity, only about one percent of central U.S. hardwood forests still remain.

In the UK the term "ancient woodland" describes woodland as an irreplaceable resource, which predates the 1600s, and in many instances is tens of hundreds of years older. Despite their heritage even these woodlands are disappearing and less than 20 percent of the total wooded areas of Britain have ancient status.

Southern Europe

One of the most important forest species in southern Europe is the cork oak that lives in mixed forest habitats alongside holm oak, zen oak, and pines. The ancient cork oak forests are now badly depleted, but they show how forests have served multiple purposes over thousands of years, supporting both people and wildlife. Great efforts are now being made to restore these old forests, which have been removed to make way for more intensive crops. In Portugal it is now illegal to cut down a cork oak unless for forest management.

Northern Asia

Although much of China is covered by desert there are nevertheless huge areas of forest. Only about 19 percent of this region still has old growth forest, yet even so it is still the location of the second largest boreal forest in the world where once the Siberian tiger roamed, though now it is limited to a tiny forest near the Sea of Japan.

The vast tracts of the "snow forests" of Asian Russia still retain areas of ancient forest which range from the Arctic zone in northeast Sakha to the southern subtropical mixed forest regions of the Ussuri and Amur basins—this is the most ecologically diverse temperate forest in the world.

Australia and New Zealand

Unique to Australia and ideally adapted to the extremes of climate and environment there are the eucalypts, a huge family of well over 700 species that account for some 70 percent of the Australian forest. Like everywhere else Australian forests are at risk from loggers, despite an attempt to limit the timber industry from felling "old growth forests." One method of protection devised in Western Australia has been to incorporate forests inside national parks.
southwest Australia also has a small area of old forest which federal law has protected for over 20 years.

In the last 100 years Tasmania has lost over seven percent of its temperate rainforests through fire, but still about ten percent of the land retains its natural temperate forests. These ancient forests contain broadleaf southern beech, pencil pine, Huon pine, and the King Billy pine

Top: Hudson Bay, near Churchill, Manitoba, Canada. This region at the Churchill River estuary on Hudson Bay is notable for its position as an "ecotone"—in other words a transitional region between different ecosystems. In this case between the taiga—or boreal forest as it is known in Canada—to the south, the Arctic tundra to the north, and the wooded Hudson Bay area. The small town of Churchill (population less than a thousand) is famous as the "Polar Bear Capital of the World."

Above: The North American porcupine (*Erethizon dorsatum*) seen in its Takshanuk Mountain homeland in southeast Alaska. Also known as the common porcupine, this primarily nocturnal animal is usually found in coniferous and mixed forests across Canada, Alaska, and most of northern and western United States. In fact a large rodent, the porcupine's ancestors probably drifted on rafts across the Atlantic from Africa to Brazil about 30 million years ago and then moved up to North America about 27 million years later.

Left: Rocky Mountain National Park, Colorado in summer after the snow has melted from all but the highest peaks, many of which are over 12,000 ft (3,658 m). This protected area comprises three distinct ecosystems: lowland meadows, evergreen fir forests—of aspen, firs, pine, and spruce—which above the tree line (over 11,000 ft = 3,353 m) becomes tundra. Thanks to this extensive range of habitat the park contains some 700 plant species and a vast range of animal life including golden eagles, black bears, marmots, moose, bighorn sheep, mountain lion, and bobcats.

Next page: Four seasonal views of an English woodland. In spring bluebells (*Hyacinthoides non-scripta*) carpet the ground as new leaves start to unfurl on the trees. By summer the bluebells have disappeared to be replaced by ferns and shade-loving woodland plants and the tree canopy cuts out most of the light bar the odd shaft of sunshine. In fall the leaves cover the ground and the trees are bare allowing lots of light into the wood. Toward the end of the year most of the plants and trees are dormant as winter brings cold weather with bitter winds, frosts, and snow.

Below: A male wolverine (*Gulo gulo*) in a forest clearing in winter. The largest land-dwelling member of the weasel family (*Mustelidae*) was traditionally trapped for its dense, water-repelling fur. The male can weigh up to 66 lb (30 kg) and be the size of a medium dog: it is also known as the skunk bear thanks to the strongly noxious odor it emits when threatened or attacked. Mainly a dweller of the high northern reaches, it also inhabits North American and North European woodlands.

Above: Snow covered conifers on the slopes near the skiing resort of Söll in the Austrian Alps. Conifers are softwood, cone-bearing trees which live in most parts of the world, but particularly in large numbers on mountain slopes. They have an ancient heritage with the fossil record placing them back to the late Carboniferous period, some 300 million years ago.

Right: View over the Norway Spruce (*Picea abies*) forest from the 65 ft (20 m) measurement tower of a meteorological station in Finland. Hidden in the deep forest, the station is located in this remote spot for its clean, pollution-free air and contributes to the global program run by the World Meteorological Organization to monitor and assess chemical composition and related physical characteristics of global background atmosphere. It is maintained by the Finnish Meteorological Institute in conjunction with the Finnish Forest Research Institute. Woodlands and forested areas are very sensitive to environmental and climate change and can offer early warning of change.

Inset: Low growing bilberry bushes cover the forest floor of a Scots pine (*Pinus sylvestris*) forest near Malung in Sweden.

Woodlands

Main image: Fall in The Mens, a large, protected, wild area of ancient deciduous woodland in the Low Weald, Sussex, England.

Above: A red squirrel (*Sciurus vulgaris*) foraging among the fall leaves of the boreal forest, Nord-Trondelag, Norway. Its favored habitats are the coniferous forests and temperate broadleaf woods of northern Europe where the best chance of seeing one is either in the morning or early evening as squirrels like to sleep during the day.

Above: A tawny owl (*Strix aluco*) flying through the light rain in the Cairngorm Mountains of Scotland. This fierce nocturnal hunter is found widely in woodlands across Europe, Asia, and parts of Northwest Africa where its favored locations are mixed and deciduous woods and forests, ideally located close to open water.

Above: An American badger (*Taxidea taxus*) peers out of a hollow log. These badgers live a mostly solitary life digging for rodents in a wide variety of habitats—particularly deciduous woodlands and open plains—from northern Mexico, across much of the western United States, and up into parts of southern Canada.

Above: Another creature of the woodland floor, the dormouse (*Muscardinus avellanarius*) is a native of northern Europe and Asia Minor. This tiny creature grows to about 3.5 in long (9 cm) and only weighs 0.7 oz (20 g), although this increases to about 1.4 oz (40 g) as it builds up fat for its winter hibernation. It feeds on fruits, berries, and nuts (especially hazelnuts) as well as insects and young leaf buds.

Above: The famous giant redwoods (*Sequoiadendron giganteum*) of Sequoia National Park, California, in the snow. Sequoias only grow from seed and can start producing immature cones when they are 12 years old: cones remain green for anything up to 20 years before they release their seeds. In (a very long) time trees can grow up to an average 280 ft (85 m) and 24 ft (8 m) in diameter. Their natural distribution area is limited to part of the western Sierra Nevada where they live in small scattered groves. The oldest dated Sequoia is 3,500 years old.

Main image: The world's largest sitka spruce (*Picea sitchensis*) trees grow in Olympic National Park, Washington State, where they reach up to 230 ft (70 m) or more tall. Seen here in July, these trees grow in the temperate rainforests that thrive in the cool Pacific breezes along the northwest coast of North America and are never found further than 50 miles (80 km) inland from the Pacific Ocean. Although sitka spruce can live for over 700 years, much of the original forest has been logged over the last century and only remnants remain.

The floral understory of a rusty gum (Angophora leiocarpa) and cypress (*Callitris glaucophylla*) woodland, with pink calytrix and many other herbs and shrubs, showing the biodiversity of native grassy woodlands in the Central Queensland Highlands, Australia.

A sugar glider (*Petaurus breviceps*) resting on branch, eating, in the Lake Eacham National Park, Queensland. The sugar glider is an arboreal (tree-dwelling) gliding possum that has a membrane stretching from its fifth finger to its first toe. Although this membrane cannot be easily seen when the sugar glider is resting, when it opens its arms it is able to glide from tree to tree for distances of up to 165 ft (50 m).

The sun sets and the moon rises over open eucalyptus woodland with gum-topped bloodwood (*Eucalyptus erythrophloia*) as it has for centuries. Although extensively cleared elsewhere, the natural cycles continue here in the Goonderoo Bush Heritage reserve, near Emerald, Central Queensland, Australia.

A koala (*Phascolarctos cinereus*) and baby on a tree trunk; the koala is described as an arboreal herbivore—a tree-living vegetarian—and despite its name is not a bear but a marsupial and the closest living relative to a wombat. Koalas like to sleep for up to 18 hours a day and can be very aggressive if disturbed. They are only found in woodlands near the coast of eastern and southern Australia where they subsist on an almost exclusive diet of eucalyptus leaves which are toxic in large quantities to other species. Uniquely, the koala is the only non-primate to have fingerprints.

Main image: The Chinese revere their lone pines: these mountain pines are in the Huangshan (Yellow) Mountains south of Anhui Province, eastern China. This extraordinarily beautiful area is a popular tourist destination and a UNESCO World Heritage Site. Often shrouded in dramatic mist, 77 mountains in the range exceed 3,280 ft (1,000 m), but the area itself is ecologically diverse: the lowest level is moist forest, then above 3,608 ft (1,100 m) it becomes deciduous forest as far as the tree line at 5,905 ft (1,800 m), when it becomes alpine grass.

Below inset: A picture of contentment, this red panda (*Ailurus fulgens*) is safely confined at the Panda Center, Wolong Valley, Wolong Tal, Himalaya. The natural habitat of this creature is the Himalayan mountains between 2,200–4,800 ft (670 m–1, 463 m), northern Burma, and parts of western Sichuan and Yunnan in China where they inhabit deciduous and coniferous forests in the temperate zone. They are solitary crepuscular animals—active at dawn and dusk and throughout the night—and sleep during the day in nests in evergreen trees. In common with their giant cousins they eat a bamboo-heavy diet but vary it with flowers and berries, small leaves, and birds' eggs.

Ice Worlds

Sub-zero temperatures and high winds are the norm at the northern and southern extremes of the Earth where thick ice caps comprise the regions known as Arctic and the Antarctic. Few life forms can survive in these challenging conditions, which are the most demanding on Earth. The Arctic surrounds the Earth's North Pole and is a vast ice-covered ocean almost entirely surrounded by permanently frozen ground, whereas the Antarctic, at the South Pole, is an ice covered continent surrounded by thick ice shelves. Despite their hostile environments, these are fragile ecosystems at the most dangerous and susceptible edge of global warming and there are palpable signs of encroaching, and possibly irreversible, change.

The Arctic

The Arctic is defined as the region of cold temperatures within the Arctic Circle at 66° 33'N. At the center lies the North Pole, the northernmost point of the Earth's axis, which is actually in the middle of the Arctic Ocean, although this area is frozen solid. The Ocean is rimmed with thousands of islands and almost completely surrounded by land, though much of this distinction is completely lost under the ice and snow. Within this area of Arctic Ocean are also parts of Canada, Greenland, Iceland, Norway, Sweden, Finland, Russia, and Alaska in the U.S. These are the lands of the midnight sun where the midsummer sun never sets and in winter the sun never appears above the horizon. Six months of continuous light followed by six months of continuous darkness.

Despite the harsh environment, various native peoples live within the Arctic Circle: the Inuit fish hunters and nomadic reindeer herders such as the Nenets, the Sami, and the Nganasan. Three much larger, settled communities also live permanently within the Arctic Circle, in Russia—at Murmansk, Norilsk, and Vorkuta—and another at Tromso in Norway is the fourth largest.

In winter, as the temperature drops to -94° F (-70° C), vast areas of the Arctic sea freezes into fractured and ridged pack ice which is continuously ripped by unforgiving winds. In summer as the atmosphere warms much of the ice melts and animal and plant life suddenly appears, including species such as dolphins and whales, bears, foxes, and many different types of sea bird. Spring flowers suddenly color the ground as plant life celebrates a brief period in the warmth of the sun. Freshwater lakes, marshes, and swamps emerge from the snows and quickly throng with migratory birds and insect life.

The animals of the Arctic have specially adapted to live and even thrive in the extreme cold. One of the most successful are the caribou (*Rangifer tarandus*) which roam in massive herds for hundreds of miles looking for food such as mosses, lichens, grasses, and dwarf willow species. Another large herbivore is the musk ox (*Ovibos moschantus*) which—until protective laws were enacted—were in danger of extinction. These animals live in small groups eating grasses, lichens, and willow leaves: happily they are slowly making a population recovery.

Arctic Hare (*Lepus arcticus*) populations fluctuate according to amounts of available food—they eat woody plants as well as buds, berries, leaves, and grass. Those hares which live at high altitude remain white all year while those inhabiting lower regions change their fur to brown in the summer for better camouflage against the dirt. Another animal which changes its coloration seasonally is the Arctic fox (*Alopex lagopus*). This beautiful creature is a scavanger that lives off abandoned carcasses left by polar bears and other large predators, though when the opportunity occurs it likes a fresh rodent, especially the lemmings that live on the tundra in small groups feeding on Arctic plants. Roughly every four years the lemming population explodes and they gather together in huge communities to range into new territories. Whenever this happens other animal species benefit from the extra food the lemmings provide: predators such as polar bears, foxes, wolverines, wolves, and snowy owls feast on the bounty.

Of course, the top Arctic predator is the omnivorous polar bear, which often hunts seals on ice floes. However, the bears are also opportunists and will feed on any meat they encounter, such as stray birds and even whale carcasses. For most of the year polar bears are solitary creatures until the mating season when they gather together to seek suitable mates. The females then disappear into snow caves for the winter where they deliver one or two cubs, which are fed on seal pups when the spring thaw reawakens the Arctic.

The Antarctic

The Antarctic lies in the southern hemisphere that overlies the South Pole and is the Earth's southernmost continent. It is entirely surrounded by the Southern Ocean which continuously circumnavigates the globe. The continent is the fifth largest in the world at 5.4 million square miles (14.4 million sq km) with a coastline of about 11,160 miles (17,968 km). It lies within the Antarctic Circle, which

Previous page: A perfect still day in the Neumeyer Channel, Antarctica, with the sea mirror-smooth and reflective. Such is the perfection of this environment that it has become a popular tourist excursion area for luxury cruise ships.

Above left: The snow-covered peaks of the Shackleton Range in northwest Antarctica. These remote mountains arise at the junction between the Filchner Ice Shelf floating on the Weddell Sea and the Antarctic Ice Sheet. Named after British explorer Sir Ernest Shackleton, the mountains run for about 99 miles (160 km) in an east-west direction.

Right: Willow ptarmigan (*Lagopus lagopus*) on the north slope of the Brooks Range along the central Arctic coast. These Arctic grouse turn white in winter and stone-color in summer so they are camouflaged against predation. They live in the circumpolar regions and eat primarily mosses, lichens, and seasonal berries.

Left: An Arctic Ground Squirrel (*Spermophilus parryii*), better known as a siksik, eating horse tail grass on the Manitoba tundra. Siksik live in colonies led by a dominant male and subsist on a varied diet of tundra plants, mushrooms, seeds, roots, and fruit. They hibernate for the winter in burrows, reappearing in the warmer spring weather.

Below right: Despite the extreme environment, the Arctic supports a fascinating variety of spring-flowering plants. The glacier buttercup (*Ranunculus glacialis*) is found above 13,000 ft (4,000 m) in the high mountains and Arctic regions. In common with its neighbors it grows low to the ground to minimize wind and snow damage.

lies at latitude 66° 33′ 39″ south of the equator. All but two percent of Antarctica is covered with ice which averages over a mile (1.6 km) in thickness, but it is also the driest continent (except around the coastal perimeter) which makes the interior—though it does not fulfill our normal expectations—the largest desert in the world. In fact, Antarctica contains about 90 percent of the ice in the world (collaterally holding some 70 percent of the fresh water) which if it melts is calculated to raise the world's oceans by 200 ft (60 m) and radically alter the world's land masses.

The only people who live here do so temporarily while on scientific or animal watching expeditions, moreover, no evidence has been found of pre-historic settlements or indigenous populations. Plant and animal life is restricted to a narrow selection of species which can endure extremes of cold for indefinite periods of time. The hostile environment has to a large extent protected Antarctica from exploitation, though the sea bed around the Ross Sea is expected to be rich in oil fields and has long been a source of contested ownership of territorial waters. Now it has been discovered that the Transantarctic Mountains contain a variety of valuable minerals including iron, chromium, platinum, molybdenum, and nickel ores. In 1959 the Antarctic Treaty was signed by 12 countries (and subsequently signed by many more) guaranteeing the prohibition of military activity and mineral mining but promoting scientific research and ecological protection. The most accessible part of the continent is the Antarctic Peninsula which has temperatures around freezing in high summer and is the home base for the many and varied scientific stations.

Antarctic is divided by a natural barrier into East Antarctica and West Antarctica by the Transantarctic Mountains (TAM for short). With a total length of just over 2,000 miles (3,500 km) these extend—with a few gaps—across the continent from Cape Adare in the north to Coasts Land in the south, in other words from the Ross Sea to the Weddell Sea. The highest mountain in the range is Mount Vinson which rises to 16,000 ft (4,897 m) in the Sentinel Range. The only places in the TAM not covered with ice or snow are a few mountain summits and some dry valleys near McMurdo Sound, and this is due to a unique Antarctic phenomenon caused by the near non-existant precipitation and erosion of ice from the surfaces.

There are many dormant volcanoes in Antarctica but the only active one is Mount Erebus on Ross Island. Another phenomenon are the more than 70 subglacial lakes which were created thousands of years ago and lie hidden deep under the ice sheet.

The most iconic animal of the Antarctic is the penguin, which is found throughout the Antarctic and sub-Antarctic regions. There are around 18 different species of penguin of which about ten live in the Antarctic. Penguins are flightless birds but superb marine animals and predators of fish, squid, and crustations. They can endure extremes of temperature and the icy environment thanks to thick layers of subcutaneous fat that enables them to breed successfully despite the inhospitable environment. The largest species is the Emperor Penguin (*Aptenodytes forsteri*), which breed in large colonies in the winter on the edge of the sheet sea ice at the edge of the Antarctic, probably the harshest breeding conditions on the planet.

Few other birds breed successfully on the Antarctic continent. In this exclusive group are the Antarctic snow petrel (*Pagodroma nivea*) and the Antarctic skua (*Catharacta maccormicki*) which breeds on the mainland between September and March.

More numerous are marine mammals of which two groups predominate—whales and seals. Both species have adapted to live in the freezing waters by evolving thick layers of blubber to insulate them from the cold, and in times of need an emergency food store. Additionally, seals have a layer of thick fur which provides them with extra insulation against the extremes of the cold when they climb out of the relatively warm water to sit on the frozen landscape and nurture their young.

Whales visit Antarctic waters in the summer to dine on the rich shoals of krill that live in abundance in the clear waters, but they return to warmer subtropical waters in the Antarctic winter to deliver their calves and only return to the Antarctic in spring when their young have grown enough to cope with the demanding journey and rough seas of the Southern Ocean.

Six species of seal inhabit the Antarctic seas and all of them are hunters who pursue squid and fish, though some eat large quantities of krill as well. The leopard seal (*Hydrurga leptonyx*) is the most dangerous and voracious predator and a rapacious pursuer of penguins and other seal species. Other Antarctic species include the solitary Ross seal (*Ommatophoca rossii*), the Weddell seal (*Leptonychotes weddellii*), the crabeater seal (*Lobodon carcinophagus*), the elephant seal (*Mirounga leonina*), and the Antarctic fur seal (*Arctocephalus gazella*). The latter are related to sea lions and live and breed principally around South Georgia where their major food source—krill of which they eat around a ton (907 kg) a day—proliferate. The strange-looking elephant seal breed in dense colonies along the Antarctic shoreline where the males fight fiercely to

keep harems of females away from other amorous males. So as not to drop their guard they do not enter the sea to feed until the risk of takeover has passed. Instead they rely on their thick reserves of blubber to tide them through.

Icebergs

Icebergs differ depending on their origin and are classified by their shape and size. Nintey percent of Arctic icebergs (some 16,000 in a typical year) calve from the glaciers along the coast of Greenland, but at least half of these are just small sections. Most of the rest calve from the Piedmont glaciers along the Gulf of Alaska. The larger pieces are caught by the currents and most drift off into the North Pacific Ocean. In the Southern Ocean the bergs sheer from the giant ice shelves along the Antarctic coast. Accordingly, they are high, and vast, tabular (flat-topped) bergs which can be over 100 miles (160 km) long and drift in the Southern Ocean. While they remain in polar waters they only diminish imperceptably during the warmer summer months and will remain stable until they reach warmer waters.

Icebergs are variously described as pinnacled, domed, blocky, tabular, arched, weathered, or valley to illustrate their shape.

Ice Sheet

There are only two ice sheets on the planet, the Antarctic ice sheet and the Greenland ice sheet. Both, in geological terms, are quite young. An ice sheet is a vast, continuous mass of glacier ice that is thousands of miles in extent. In previous ice ages other great ice sheets covered parts of the planet. The Antarctic ice sheet covers an area of almost 5.5 million square miles (14 million sq km) and is the largest single mass of ice on Earth. In East Antarctica the ice rests on a major land mass, however in West Antarctica the land surface is still covered in ice but down to 8,000 ft (2,500 m) below sea level—if this were not frozen, it would be the sea bed.

The Greenland ice sheet covers around 83 percent of the land, the only non-frozen areas being around the coastline. If this were to melt it is estimated that sea levels would rise by over 23 ft (7.2 m).

Far left: Orange and yellow lichen encrusting rocks in Antarctica. Caribou and musk ox are some of the very few animals adapted to eat lichen. A unique organism in that they are a symbiosis of fungus and algae, lichens are very sensitive to air pollution and only grow where the air is pure. They can flourish in all types of environment, even extreme icy conditions, and are estimated to cover eight percent of the Earth's surface.

Left: A summer sunset over sea ice on the Weddell Sea. Named after the British navigator James Weddell who discovered it in 1823, the sea is that part of the vast Southern Ocean where it meets the Antarctic peninsula. Much of the sea is part of the Filchner-Ronne Ice Shelf and is permanently frozen, although in recent years the shelf has started melting rapidly.

Glaciers

Although rarely seen by most people, glaciers cover as much as ten percent of the world's surface. They are slow moving "rivers" of ice which are so heavy that they achieve momentum. Glaciers form when the temperature is so low that snow and ice accumulate for many years with little or no summer melt. The crystalline microstructure of snow traps air, so the ice does not become a solid dense mass; untouched snow remains light with a density of around 110 - 661 lb per cubic foot (50 - 300 kg per cu m). As further snow falls accumulate, the underlying snow is compressed and some portions melt to become a formation of crystals known as "firn." By then they have a density of around over 1,000 lb per cubic foot (500 kg per cu m). As the weight accumulates the lower layers are compressed further into solid blue ice—this can take anything from five to 3,000 years. When the build up of ice is thick enough the glacier starts to form but it will not be subjected to sufficient pressure to move until it is around 164 ft (50 m) thick.

The head of the glacier (known as the "source") collects new ice and in the right conditions will "push" layers of blue ice and firn so the glacier advances across the landscape. At the front or "snout" the ice will break off (or melt in warm conditions) when it meets water and large pieces become icebergs when they break off (calved).

Permafrost

The term "permafrost" describes land that remains permanently frozen, even during the summer months. Permafrost layers can remain underneath non-frozen soil and present an impermeable barrier that prevents rainwater from draining away. No trees grow in such soils and few plant species other than reindeer moss survive.

Permafrost lands which have been solid for thousands of years are now starting to thaw and the trapped carbons they contain—around a third of the soil-bound carbon in the world—are released as carbon dioxide into the atmosphere, with dire consequences for the planet.

Nunataks

Nunataks are exposed and isolated mountain peaks surrounded by ice. They stick out by virtue of their lack of snow cover and tower a great height above their level surroundings. They are frequently named as they provide invaluable reference points for travelers in an otherwise featureless landscape. The lifeforms found on nunataks are often unique as they have developed in isolation from all other species.

Main image: Drift ice on the Arctic Ocean at Svalbard, Spitzbergen, Norway. Drift ice tends to be seasonal and melts in the summer. The Svalbard archipelago lies north of continental Europe and far north of the Arctic Circle within the Arctic Ocean and consists of a group of islands, of which the three largest are populated: Spitsbergen, Bear Island, and Hopen.

Far left below: A female polar bear (*Ursus maritimus*) playing with her newborn cubs outside their den in the eastern Arctic National Wildlife Refuge at the mouth of the Canning River in Alaska. Polar bears are the largest land predators in the world and have superbly evolved to live within a very narrow and extreme cold environment, namely the Arctic Ocean and its surrounding seas.

Below middle: A barren ground caribou (*Rangifer tarandus*) herd moving along the Haul Road of North Slope, Alaska, in summer. These animals have yet to grow their mature antlers which reach full size in September and then shed within a few months after the rut. Both male and female caribou grow antlers, the size of which is linked to how well they are feeding.

Below: A musk ox (*Ovibos moschantus*) cow with her newborn calf seen on the north slope of the Brooks Range on the central Arctic coastal plain of Alaska. These animals gather in herds of up to 20. When danger is sensed they form an outward facing defensive circle around their young. Their diet consists of grasses, lichens, and willow leaves.

Above: Snowshoe hare (*Lepus americanus*) range across Canada and in the northernmost U.S. along the Sierras, Rockies, and Appalachian mountain ranges. In summer their fur is predominantly gray-brown, while in winter to camouflage into the snow they turn completely white except for black ear-tips and eyelids.

An Arctic fox (*Alopex lagopus*) caught in mid-spring while hunting rodents in Alaska. Arctic foxes live in burrows and can survive extreme temperatures down to -58° F (-50° C). In summer their coat turns brown or gray according to their environment, while for winter they turn snowy white which makes them near invisible. They live on a diet of rodents, birds, and fish.

Previous page: Global warming has led to the earlier melt of the ice pack, which often results in stranding polar bears. Although they are strong swimmers and can usually easily escape, the early melt makes it difficult for the bears to hunt and they can soon become weak, starve, and even drown.

A rock ptarmigan (*Lagopus mutus*) in the Arctic National Wildlife Refuge, Alaska. This vast refuge encompasses five different ecological zones—coastal marine, coastal plain tundra, alpine tundra, forest tundra transition, and boreal forest—spanning around 200 miles (322 km) north to south. Of particular note are the polar bears, caribou herds, grizzly bears, wolves, snow geese, and peregrine falcons.

The Arctic ground squirrel lives in small groups of related females associated with a dominant male. They live in burrows dug into sandy and gravel areas with good drainage, such as river banks, alongside lake shores, and on moraines and eskers in Alaska and northern Canada. Primarily a herbivore, the Arctic ground squirrel is believed to be the only animal able to lower its body temperature to below freezing to help it survive the long Arctic winter.

Above: A breeding colony of Adelie penguins. In the surrounding waters lurk leopard seals and killer whales waiting for their unlucky prey. Penguins are not easy victims as they can swim very fast and dive down to 500 ft (152 m) if necessary. The penguins in turn live on fish and krill but their populations have more than halved in the last 25 years due to scarcity of food and the reduction of sea ice.

Left: Adelie Penguins (*Pygoscelis adaliae*) diving off the edge of the ice to collect food for their waiting chicks. This is one of the most common and smallest penguin species and Adelie penguins can be found in their thousands on headlands, beaches, and islands all along the Antarctic coast. They spend the winter at sea living in groups on the pack ice and icebergs but return to the stony shores of the Antarctic to breed in huge rookeries in September and October.

Right: A leopard seal (*Hydrurga leptonyx*) waiting on an ice flow near Palmer Base, Antarctica. This top Antarctic predator is a solitary animal except during the breeding season and is an endangered species. It lives on a diet of penguins, squid, and krill and in turn is only predated by killer whales.

Previous page: Pack ice floating in the Arctic Ocean around Spitzbergen, Svalbard. Pack ice is formed at sea from frozen sea water and drifts with the ocean currents and winds. It can be very unstable as it continually cracks, breaks and refreezes and can be anything from a few inches to many feet thick. In winter the ice cap expands to cover roughly eight percent of the southern oceans and five percent of northern seas where it makes up much of the ice cap in the Arctic Ocean. Pack ice is a considerable hazard to shipping.

Previous page: Emperor Penguins (*Aptenodytes forsteri*) and chicks huddling together for warmth and protection during a blizzard on the Dawson-Lambton Glacier in Antarctica. Between 5,000 and 10,000 brooding pairs make up the colony. At the end of fall each female lays a single egg then disappears off to sea leaving the male to starve for two months while he balances the lone egg on his feet. The females reappear when the young hatch and the males return to sea to feed and regain strength. At four months the parents leave them and a month later the young penguins themselves go to sea to feed and fend for themselves.

Next page: A giant slice of ice falls from the face of Perito Moreno Glacier in Los Glaciares National Park, Argentina. Perito Moreno Glacier is around three miles (5 km) wide at its sea base end and towers 197 ft (60 m) above the surface of the water, under which it plunges to around 558 ft (170 m). The glacier advances at a rate of about 6.5 ft (2 m) a day, (or just under half a mile a year), but it also loses ice at much the same rate, so the ice front has barely changed in the last century.

Above: Icebergs are large chunks of glacial ice which have broken off from their ice shelf or parent glaciers. Two thirds of the berg lies below the water. While the iceberg remains in polar waters its lifespan is indefinite, but once in warmer waters it starts to melt. Icebergs originating in Antarctica break off the giant ice shelfs that rim the continent and then drift in the Southern Ocean. These are typically flat topped and can be several hundred feet high and over 100 miles (160 km) long.

Left: The steep sided Lemair Channel cuts between the mountains of Booth Island and the Antarctic Peninsula for 7 miles (11 km). This has become a prime tourist destination thanks to its clear, glassy still waters, though the channel is ice bound in winter and early spring. At the northern end of the channel are the two rounded snow cap peaks of Cape Renard on Renard Island.

Right: As the iceberg melts it assumes sculptural shapes and strange bluey colors. The frozen ice from glaciers can be tens of hundreds of years old.

Rainforests

Generally speaking rainforests are luxuriant forests—dominated by tall trees—that receive a high degree of rainfall, averaging in the region of 70 to 100 inches (about 1,800 to 2,500 mm) per annum. Despite occupying only between six and seven per cent of the Earth's surface, rainforests are of huge environmental importance; the rainforests have been described as the "lungs of the planet" because their lush vegetation continuously recycles carbon dioxide, providing much needed oxygen. The rainforests are also world's most biologically diverse ecosystems, containing an estimated fifty per cent of the world's known plant and animal species. Just consider, the total number of species of tree that can be found in all of the forests of the United States and Canada is around 700—in an area considerably less than one square mile (2.6 sq km) of a Malaysian rainforest more than 375 species were found.

Types of Rainforest

When most people think of a rainforest they imagine a hot, tropical environment but, in fact, there are two types of rainforest: tropical and temperate. Of these two, tropical rainforests cover a much greater portion of the Earth's surface.

Tropical rainforests are generally to be found between the Tropic of Capricorn and the Tropic of Cancer (23°26′ 22″ south and 23°26′ 22″ north of the equator respectively) in parts of Central and South America, Central and Western Africa, and Southeast Asia and Australia. Because of their close proximity to the equator the climate in tropical rainforests remains fairly constant throughout the year. The weather is consistently warm and wet; rainfall is spread out evenly over the year, the average temperature being between 20º C and 29° C (68º F and 84° F) and rarely falling below 18° C (64° F).

Temperate rainforests are generally to be found further north and south of the equator than tropical rainforests, in parts of Western North America, Southwestern South America, Western Europe, East Asia, Australia, and New Zealand—in most cases they are located relatively close to the sea. Although they may receive a similarly high amount of rainfall to tropical rainforests—50 to 60 inches (about 1,270 to 1,525 mm) though this is exceeded in some cases—temperate rainforests are more seasonal in nature. They are, however, far less seasonal than other regions in similar areas, tending to have milder, wet winters and cooler summers. Summer fogs and high levels of cloud cover are common in many temperate rainforests during the hottest months and these help to keep them cool and moist.

Previous page: A view looking up through a canopy of fan palms, fronds, and green leaves. The leaves and branches of the taller trees of the emergent layer can be seen in the sunlight above.

Right: A lush subtropical rainforest of piccabeen palm (*Archontophoenix cunninghamiana*) at Woongoolbver Creek, Fraser Island World Heritage Area, Queensland, Australia. Although made entirely of marine sands, Fraser Island supports dense vegetation that recycles its own dead matter for nutrients.

Far right above: The Rafflesia (*Rafflesia pricei*) is a parasitic flowering plant, found in Indonesia and named after the leader of the expedition that first discovered it—Sir Thomas Stamford Raffles. The flower can reach up to 39 inches (100 cm) in diameter and weighs up to 22 lbs (10 kg). The Rafflesia gives off a strong stench, like that of rotting meat, which it uses to attract carrion flies.

The Layers of the Tropical Rainforest

Tropical rainforests can be divided into four distinct layers, extending down from the tops of the highest trees to the ground below. Each layer is home to a wide variety of different plants and animals that have adapted to life at that particular level.

The *emergent layer* is occupied by the tops of the very tallest trees and these treetops are not generally very close to one another. The trees of the emergent layer grow to heights of over 200 ft (61 m) and, due to their exposure to the elements, are subject to the most extreme weather conditions. They receive more direct sunlight and less moisture than the trees in the layers below and are also exposed to strong winds and heavy rain. These higher reaches of the rainforest are home to butterflies, bats, eagles, small monkeys, and insects.

The *canopy layer* is home to the largest range of biodiversity in plant and animal life in the rainforest. Here trees grow to heights of 100 to 150 ft (30 to 45 m) and their branches are very closely packed, forming an almost continuous covering of foliage—hence the name "canopy." This dense layer has a huge impact on those above and below it for it traps moisture and humidity beneath the overlapping leaves, thereby stopping them from reaching the emergent layer and also affords protection from sunlight and heavy rainfall to the layers below. It is estimated that over 50 percent of the plant and animal species of the rainforest live within the canopy layer.

The *understory layer* lies between the canopy and the forest floor. At this level very little sunlight penetrates—less than five per cent—and both trees and plants have evolved with larger leaves to try and capture as much of the available sunlight as possible. The younger trees in the understory tend to grow no higher than 66 ft (20 m) tall, they must wait for one of the older trees overshadowing them to die, thus creating a gap in the canopy overhead and granting access to the sunlight required for further growth. The dark and humid

Left: The glasswing butterfly (*Greta Oto*) is a brush-footed butterfly found mainly in Mexico and Central America and inhabiting the rainforest understory. Its name is derived from the fact that the tissue between the veins of its wings is transparent.

Right: Originally from South America but now widespread through West and East Africa and Southeast Asia, the kapok tree (*Bombax rhodognaphalon*), which is also known as the silk-cotton tree, grows to heights of over 165 ft (50 m) and often forms part of the emergent layer of rainforests. The seeds, leaves, bark, and resins of the kapok have been used to treat dysentery, fever, asthma, and kidney disease.

environment of the understory makes it the ideal habitat for the large number of insects that live there, alongside birds, butterflies, frogs, lizards, and snakes.

The *forest floor* receives even less light than the understory; as little as two percent. Due to this lack of sunlight, the forest floor has quite a sparse covering of plant-life, although it does contain a layer of decomposing organic matter that decays more quickly than usual because of the warm, humid conditions. This decaying matter is consumed by a variety of detritivores—animals and plants that feed on such matter—including millipedes, dung beetles, woodlice, and many species of fungi. The rapid decomposition of this layer of decaying organic matter means that the soil quality of the forest layer is frequently quite poor. This, as much as the lack of sunlight, is responsible for the sparse plant-life on the forest floor.

The Destruction of the Rainforests

It has been estimated that as recently as the 19th century as much as 20 percent of the Earth's landmass was covered by tropical forests. By the end of the 20th century this had dropped to below seven percent.

One of the major causes of this deforestation is logging, and as technology has advanced so it has enabled increasingly large areas of forest to be cut down in increasingly short periods of time using less manpower. Left alone after loggers have finished, some rainforests will regenerate given time, but—unfortunately—this is frequently not the case. With the increase in population, the demand for land has meant that in many instances the areas abandoned by loggers are then occupied by farmers who clear them to sow crops. As the soil in most rainforest areas is poor, the farmers soon realize that it will not yield a sufficiently good harvest and sell the land on to cattle farmers. The area is then lost to forestation as it either stays as grazing land for the cattle or is left in such a desolate state that it is virtually impossible for the rainforest to regenerate without human intervention.

In the late 20th and early 21st centuries there has been a much greater global awareness of the ecological significance of the world's rainforests and many organizations, alongside some governments, are now—thankfully—making a concerted effort to dramatically decrease the levels of deforestation.

South American Rainforest: The Amazon

The Amazon Rainforest is without doubt the most famous in the world, it is also the largest—stretching from the Andes Mountains in the west of South America to the Atlantic Ocean on the east. The Amazon Rainforest occupies portions of nine countries—principally Brazil, but also Peru, Colombia, Venezuela, Ecuador, Bolivia, Guyana, Suriname, and French Guiana—and covers a total area of some 2.3 million square miles (6 million sq km) while at its broadest it reaches a width of 1,200 miles (1,900 km).

Flowing from one side of the rainforest to the other (from west to east) is the mighty Amazon River—stretching some 4,000 miles (6,400 km) from its source in the Andes to its mouth in the Atlantic. During the high water season the Amazon discharges an incredible 6,180,000 cubic feet (175,000 cubic m) of water per second into the Atlantic—enough that it dilutes the salinity of the seawater for over 100 miles out into the ocean, so far that in the past sailors were able to drink fresh water here before they even sighted land.

The Amazon Rainforest is a treasure trove of biodiversity, home to millions of species of plants, birds, insects, and a myriad of other creatures, many of them—such as the Maues marmoset, that was only discovered in the 1990s and the entire population of which lives in within an area of just few square miles—unique to the region.

African Rainforest

Occupying large areas of Western and Central Africa, the rainforest of the Congo basin is the second largest on the planet—only the Amazon is larger. Like the Amazon it has been greatly

depleted during the last century—some estimates claim that as much as 90 percent of the rainforest has been lost in the last hundred years.

The vast majority of the African Rainforest is located in the Democratic Republic of Congo (DRC), but it also covers large areas of Gabon, Cameroon, Equatorial Guinea, the Central African Republic, and the Republic of Congo.

The major river that flows through the African Rainforest is the Congo, which stretches for around 2,900 miles (4,700 km) from its source in the highlands of northeastern Zambia to its mouth in the Atlantic Ocean—located in the DRC. Although its highest rate of discharge into the ocean—1,450,000 cubic feet (41,000 cubic meters) per second is considerably less than that of the Amazon it is nonetheless the second largest in the world.

Three species of great ape can be found in the African Rainforest—gorillas, chimpanzees, and bonobos—as well as a number of unique species, such as the Congo peacock and the okapi (a plant eating mammal, with striped legs, that looks like a small giraffe without the long neck).

Asian Rainforest: Sumatra

Located on the island of Sumatra—in the Indian Ocean—Tropical Rainforest Heritage of Sumatra was listed as a United Nations Educational, Scientific and Cultural Organisation (UNESCO) World Heritage site in 2004. It is made up of three Indonesian national parks: Gunung Leuser National Park, Kerinci Seblat National Park, and Bukit Barisan Selatan National Park. The total area covered by the three parks is around 9,650 square miles (25,000 sq km).

The parks contain an abundance of wildlife and fauna, some 10,000 species of plants, more than 200 species of mammals, and around 580 bird species. These include 21 species of birds that are unique to the area as well as the Sumatran orangutan that is, likewise, only to be found here.

Far left: A female orangutan (*Pongo pygmaeus*) with her youngster. Orangutans are not as large as gorillas, but are generally bigger than chimpanzees. The adult male of the species is usually twice as large as the female, weighing up to 285 lb (130 kg) and standing up to 4.3 ft (1.3 m) tall.

Left: A view of the canopy and emergent layers of the lowland dipterocarp rainforest near Mount Kinabalu, Sabah, Malaysia. Mount Kinabalu is the tallest peak in the Malay Archipelago, reaching 13,455 ft (4,101 m).

Next page: A panoramic view of a section of the the Iguazu Falls in Iguacu National Park, on border of Brazil and Argentina. At 1.7 miles (2.7 km) wide, the Iguazu Falls are almost three times the width of Niagra Falls.

Australian Rainforest: Daintree Rainforest

The east coast of Australia is also home to a UNESCO World Heritage site, the Gondwana Rainforests of Australia. Gondwana was an ancient supercontinent that began to break up some 180 million years ago and was comprised of modern day Africa, Australia, Antarctica, Arabia, India, Madagascar, and South America. Daintree Rainforest, north of Cairns in North Queensland, is the largest of the forests that make up the Gondwana Rainforests of Australia—covering an area of around 745 square miles (1,200 sq km).

The Daintree Rainforest is home to a significant amount of Australian wildlife; 65 percent of Australia's bat and butterfly species can be found here, along with 30 percent of frog, marsupial, and reptile species, and 20 percent of bird species (over 400 species of birds, of which 13 are unique to this area).

Top left: Leafcutter ants (*Atta cephalotes*) have very powerful jaws that they use to slice off pieces of leaves that are then carried back to their underground nest. Amazingly they are able to carry up to 20 times their own body weight. Social insects, their colonies can contain many millions of ants and columns of leafcutters transporting food can measure as long as 100 ft (30 m).

Middle left: Poison dart frogs (*Dendrobatidae*), as their name suggests, secrete poison from their skin. They are native to South and Central America and some South American tribes use the poison to coat the tips of arrows. Most species have brightly colored skin that acts as a deterrent to predators.

Below left: Red-eyed tree frogs (*Agalychnis callidryas*) are nocturnal amphibians that live in lowland tropical rainforests in Central America and northern South America. They tend to live near to rivers, as this is where they go to lay their eggs. Although they mainly eat insects it is not unknown for red-eyed tree frogs to eat other, smaller frogs.

Below: The blue morpho butterfly (*Morpho menelaus*) is notable for its iridescent wings. As seen in this picture the underside of the blue morpho's wings are colored brown but when it opens its wings you can see the brilliant blue top side that changes color depending on the angle at which they are viewed.

Above: The spider monkey (*Ateles geoffroyi*) has very long spindly limbs, hence its name, and a tail that is actually longer than its body. They can be found in the highest branches of the rainforests of southern Mexico and Brazil, living in bands of up to 35 members.

Right: The white-throated toucan (*Ramphastos tucanus*) is found in tropical South America east from Colombia and Bolivia to southern and eastern Brazil. They generally grow to a size of 21 to 24 in (55 to 60 cm) in height and their brightly colored, large bill is typically 5.5 to 7 in (14 to 18 cm) in length.

Above: The buff-tailed coronet (*Boissonneaua flavescens*) is a species of hummingbird found in Colombia, Ecuador, and Venezuela. Unlike other birds, hummingbirds' wings are only connected to their body at the shoulder joint; this enables them to fly not only forward, but also backward, sideways, vertically up and down, and to hover in front of flowers as they feed.

Left: Angel Falls in the Guiana Highlands in Bolívar state, southeastern Venezuela is the highest waterfall in the world, dropping a distance of 3,212 ft (979 m). Discovered in 1935, the falls were named after the American adventurer James "Jimmie" Angel.

Below: An array of chit palms (*Thrinax radiata*) in the understory of the rainforest in the Tikal National Park in Petén, Guatemala. Tikal National Park was established in the 1950s and was listed as a UNESCO World Heritage site in 1979.

Above: A male adult chimpanzee (*Pan troglodytes*) sitting on the branch of a tree in the Gombe Stream National Park in Tanzania. This park—covering a 20 square mile (52 sq km) strip of land near the shore of Lake Tanganyika—was formed in 1960s specifically to protect the chimpanzees.

Left: A female mountain gorilla (*gorilla berengei*) and two young in the Volcanoes National Park (Parc National des Volcans) in Rwanda. More than half of the world's remaining mountain gorillas can be found here. The park, and its gorillas, were made world-famous by the movie "Gorillas in the Mist."

Right: The bamboo lemur (*Hapalemur griseus griseus*), also known as the gentle lemur, can only be found on the island of Madagascar, mainly in the eastern rainforest areas that contain vast amounts of bamboo.

Main image: A view of trees and foliage beside Baffle Creek in central Queensland, Australia. The plant life provides essential stream bank protection and aquatic habitat microclimate control. Baffle Creek system is the major coastal river system in Queensland.

Above right inset: Buttresses of sinuous old-growth subtropical rainforest tree in a grove of piccabeen palm (*Archontophoenix cunninghamiana*) in The Palms National Park, southeast Queensland, Australia. Rainforest trees are typically shallow rooted to harness surface nutrients from decomposing leaf litter, as deeper soils are often leeched of nutrients by the heavy rains.

Right inset: Deadly vines enshroud a grand old burdekin plum tree (*Pleiogynium timorense*) and are on the way to causing its slow death by strangulation, in forest near Miriam Vale, central Queensland, Australia. The culprit is the introduced cat's claw creeper, perhaps the worst of a suite of vine weeds that are devastating or threatening old-growth trees in subtropical and tropical Australia as well as elsewhere.

Above: A view of lowland tropical peat swamp forest, near the Sekonyer River, in Kalimantan, Borneo. Peat swamps store large quantities of water and release them slowly into the surrounding area, thus ensuring a constant supply.

Above right inset: A young Balinese macaque (*Macaca fascicularis*) feeding on a nut, in Sacred Monkey Forest, Ubud, Bali. Female macaques tend to remain with the troop into which they are born, whereas male macaques frequently move between troops.

Left: An aerial view of tropical rainforest in Borneo, with the river meander almost forming an oxbow lake. The dark brown color of the river is caused by the build up of silt in the water. There is no seasonal leaf loss in this equatorial tropical rainforest, but individual trees flower and drop leaves at different times of year.

Right: A Thomas leaf monkey (*Presbytis thomasi*) in Gunung Leuser National Park, Bukit Lawang, Sumatra. Only to be found in Indonesia, the Thomas leaf monkey's rarity has led to it being put on the International Union for Conservation of Nature (IUCN) Red List of threatened species

Above: Among the most beautiful of the creatures that live in the rainforest are the brightly colored varieties of birds of paradise. Mainly to be found in New Guinea and the nearby islands as well as in Australia, most of the 40 or so species of the birds of paradise feed on insects, spiders, and fruits.

Main image: A lowland equatorial freshwater swamp forest during the wet season in Way Kambas National Park in Lampung province, southern Sumatra, Indonesia. The park is famous for its elephants but is also home to a few Sumatran tigers.

Top: The Malayan sun bear (*Helarctos malayanus*) is the smallest member of the bear family, generally growing 3.3 - 4 ft (1 -1.2 m) tall and weighing between 59 and 143 lb (27 - 65 kg). They are mainly nocturnal and feed on insects, fruit, honey, and small vertebrates such as lizards and birds.

Above: The distinctively marked clouded leopard (*Neofelis nebulosa*) has large upper canine teeth that are longer than those of any other living feline. These medium-sized cats also have an extremely long tail—sometimes as long as their body—that is marked with black rings.

Left: Cymbidium orchids—also known as boat orchids—pictured in the Doi Inthanon National Park in Chiang Mai Province, Thailand. As well as growing in soil these flowers can be epiphytic (able to grow on another plant—as seen here, growing on a tree) or even lithophytic (able to grow on rocky or stony ground).

Right: An array of dancing doll orchids (*Oncidium pusillum*) growing on the branches of a rainforest tree. These flowers thrive in the shady and damp conditions of the forest canopy.

Below: The black orchid (*Encyclia Cochleatum*) is the national flower of Belize. This orchid grows is epiphytic and tends to grow on trees in damp areas. It flowers nearly all year round and its clustered bulblike stems can grow up to six inches (15 cm) long.

Below right: Dendrobium orchids (also known as Singapore orchids)—seen here growing on the trunk of a rainforest tree—take their name from the Greek words "dendron," meaning "tree" and "bios," which means "life," referring to where the plant is often seen growing.

Great Plains

Although the names used to describe them vary—tundra, pampas, steppe, prairie, and savanna—great plains can be found on every continent of the world besides Antarctica. These huge swathes of open land cover almost a quarter of the Earth's land surface.

Whatever name they are given, there are two things that distinguish the great plains of the world: they occur where the level of annual rainfall is too low for forests to survive and too high for deserts to exist—generally somewhere between 10 and 30 inches (255 mm to 760 mm) per annum—and their weather is extremely seasonal with harsh winter months and wet and/or temperate summers that allow for an explosion of plant life.

Seas of Grass

Come the summer months and their life-giving rains, the great plains are dominated by grasses of many kinds that provide food for the vast herds of animals that arrive at this time of year. There are thousands of different species of grass throughout the world and those that cover the plains are among the most hardy and versatile of plants, able to survive extremes of heat and cold, flooding, intense grazing, and even, in some cases, burning.

The secret of these grasses ability to prosper in these extreme circumstances is that that they are able to grow very quickly when favorable conditions arrive and that, unlike most other plants, their leaves grow up continuously from the base of the plant, thus enabling them to constantly regenerate.

The North American Great Plains

The Great Plains of North America are an enormous, generally high, plateau of semiarid grassland, stretching for around 3,000 miles (4,800 km) from the three Prairie Provinces of Canada—Manitoba, Saskatchewan, and Alberta—down through ten states of the United States—Montana, North Dakota, South Dakota, Wyoming, Nebraska,

Kansas, Colorado, Oklahoma, Texas, and New Mexico. Bordered by the Mackenzie River to the north, the Canadian Shield to the east, the Rio Grande to the south, and the Rocky Mountains to the west the Great Plains (more commonly referred to as "the Prairies" in Canada) vary in width from around 300 to 700 miles (483 km to 1,127 km) and cover an overall area of some 1,125,000 square miles (2,900,000 sq km).

The major rivers that drain the Great Plains are the Missouri, Red, Rio Grande, and Arkansas in the United States and the Saskatchewan in Canada. Annual rainfall in the area generally increases as you travel from west to east and this means that the grass types that are found vary accordingly—shortgrass in the rain shadow of the Rocky Mountains to the west, mixed-grass in the central Great Plains, and tallgrass in the wetter regions to the east.

The land of the Great Plains is by and large flat although some small hills and mountains in the Ozark Plateau of Missouri, and in the Boston and Ouachita Mountains of Arkansas and Oklahoma punctuate this flat landscape.

Previous page: A herd of wildebeest (*Connochaetes taurinus*), seen here in single file, during the annual migration across the Maasai Mara. Acacia trees can be seen dotted across the plains as they rise into the hills in the background.

Above left: A view of the shortgrass prairies grassland in Pawnee National Grasslands, near Fort Collins, Colorado. The dominant species of shortgrass that can be seen are buffalo grass (*Buchloe dactyloides*) and blue grama (*Bouteloua gracilis*).

Right: A close up of the head of a bison (*Bison bison*), seen here crossing the Firehole River in Yellowstone National Park, Wyoming. Although both male and female bison have the short, upcurved horns that can be seen here, those of the male are generally larger and are used during fighting.

Animals of the North American Great Plains

The Great Plains is home to a great variety of wildlife, over 1,500 species of plants, 300 species of birds, almost 100 mammals, and over 200 species of butterflies.

The *American Bison* (*Bison bison*) is more commonly known as the buffalo or the plains buffalo and is perhaps the most iconic of the animals that roam the Great Plains. It has been estimated that prior to the arrival of European settlers in the United States the bison population was in excess of 50 million, estimates put the modern population at somewhere in the region of just 500,000. This is, however a vast

Left: A solitary male Asiatic wild ass *(Equus hemionus)*—also known as the onager—walking on the dusty plains of the Little Rann of Kutch National Park in Rajasthan, India. These horse-like animals used to be domesticated, but are now thought to be in danger of extinction.

Below right: A panoramic view of the arid Mongolian steppes, in the shadow of the foothills of the Himalayan Mountains. A herd of yaks (*Bos grunniens*) can be seen grazing in the foreground.

improvement from the latter part of the 19th century when the bison faced total extinction due to unremitting slaughter at the hands of the white settlers—one of the major causes of conflict between the settlers and the Native American Indians. At its lowest point the bison population in the United States fell below 1,000.

Although the majority of bison have a coat of thick, shaggy dark brown fur very occasionally a bison is born with white fur—these white bison are treated as sacred in some Native American religions.

Male bison are generally larger than the females with a fully-grown bull weighing up to 2,000 lbs (907 kg) and standing as tall as 6.5 ft (2 m) at the shoulder as compared to a fully-grown female that weighs up to 700 lbs (320 kg) and stands up to 5 ft (1.5 m) tall at the shoulder.

The metabolism of the bison is well adapted to its environment; in the harsher winter months when food is much more scarce it slows down to roughly a quarter of the summer rate.

The *black-tailed prairie dog (Cynomys ludovicianus)* is another animal whose numbers have declined sharply over the past couple of centuries. However, as of 2004 they have been removed from the U.S. Fish and Wildlife Service Endangered Species Act Candidate Species List.

The black-tailed prairie dog—along with four other species of prairie dog: the white-tailed prairie dog, the Gunnison prairie dog, the Utah prairie dog, and the Mexican prairie dog—are rodents of the squirrel family that can only be found in North America. The average length of an adult, including tail, is between 14 and 17 inches (36 cm to 43 cm) and they weigh between one and three pounds (0.5 kg 1.5 kg).

Black-tailed prairie dogs are burrowing animals that create extensive and often elaborate burrows with numerous entrances. Unlike the white-tailed, Gunnison, and Utah prairie dogs they do not hibernate, as with the bison their metabolism has adapted to enable them survive times of scarce food in the winter.

The *pronghorn* (*Antilocapra americana*) is also known as the prongbuck and the pronghorned, or American, antelope and is similar in appearance to the African gazelle. It is the second fastest land animal in the world—only the cheetah can run faster—and is the fastest in the Western Hemisphere, reaching speeds of over 55 miles per hour (8460 kph). They can, however maintain these high speeds for far longer than the cheetah is able to. As well as being very fleet of foot pronghorns are able to jump up to 20 ft (6 m) in one leap.

Both male and female pronghorns grow horns—the females being generally shorter—on the top of their heads. These horns are made up of a bony core that is covered by a hair-like growth that is shed annually.

As with the bison, the pronghorn population has shrunk form a high in the region of 50 million to somewhere around 500,000.

African Great Plains: The Serengeti

Probably the most famous of the great plains of Africa is the Serengeti in Kenya and Tanzania that takes its name from the Maasai word *Siringitu* that roughly translates as "the place where the land goes on forever." The Serengeti covers an area of some 11,500 square miles (30,000 sq km) and contains a number of national parks and game reserves; the Serengeti National Park, the Ngorongoro Conservation Area, Maswa Game Reserve, the Loliondo, Grumeti and Ikorongo Controlled Areas and the Maasai Mara National Reserve in Kenya. The Serengeti National Park and the Ngorongoro Conservation Area are both UNESCO World Heritage Sites.

Animals of the Serengeti

The *wildebeest* (*Connochaetes taurinus*) is also known as the gnu and is part of the antelope family, though it resembles a cow with an ox-like head. Adult wildebeest reach a height of between three to four feet (1 to 1.3 m) at the shoulder and weigh 330–550 lbs (150–250 kg).

The Serengeti is probably most famous for the extraordinary annual migration of the wildebeest—alongside zebras and gazelles. From December to March the wildebeest can be found on the grassy plains below the Ngorongoro Crater in Tanzania—it is here that the calving season occurs. When the rains here end the plains begin to dry out, turning from fertile grassland into arid semi-desert. At this point, usually in late March/early April, over a million wildebeest and thousands of zebras follow the rains to the north and west, toward LakeVictoria.

By the end of May the huge herds of wildebeest are once again on the move, this time heading further north—the mating season occurs during May and June. By July the wildebeest are massed along the banks of the Mara River that must be crossed to reach the now-fertile Maasai Mara, where the herds stay through until November, when they begin to head back south.

The *African savanna elephant (Loxodonta africana oxyotis)* is also known as the African bush elephant and is the largest terrestrial mammal. The males of the species range in length from 19 to 20 ft (5.8 m to 6 m), stand between 10 and 12 ft (3 m to 3.6 m) at the shoulder, and weigh an average of 8,000 lbs (3,630 kg). Herds of African savanna elephants are made up of related females and their young and are presided over by the eldest female (the matriarch). Adult males tend to live solitary lives and only come to the herd during the mating season. The gestation period of the African savanna elephant—22 months—is the longest among mammals.

The Arctic Tundra

Sitting at the top of the world, the Arctic tundra lies north of the coniferous forest belt. Similar areas that occur above the tree line of high mountains are known as Alpine tundra. The climate of the Arctic tundra is characterized by a long, dark, frozen winter with little rain or snowfall followed by a short summer. But when the summer does arrive it transforms the barren landscape, as the snow and ice begin to melt the first shoots of grass start to emerge and, within a few days there is a plentiful supply of food available for the millions of birds and animals that migrate to the Arctic, having spent the summer in warmer, more hospitable climes to the south.

Animals of the Arctic Tundra

Caribou (*Rangifer tarandus*) spend the Arctic winter in the forests to the south and migrate north through Alaska and Canada in huge herds of up to 500,000 during springtime. The annual round trip of around 3,000 miles (4,828 km) that the caribou make during their migration is the longest of all land animals.

Caribou are both quick runners—reaching speeds of up to 50 mph (80kph)—and very strong swimmers. They have lightweight, hollow hair that

Left: A gerenuk *(Litocranius walleri)* standing on its hind legs to feed from the remaining green leaves at the top of a bush in the Samburu National Reserve on the banks of the Ewaso Ngiro River in Kenya. The gerenuk is a slender member of the antelope family that takes its name from the Somali for "giraffe-necked."

Right: A herd of Burchella zebra *(Equus buchelli)* grazing on the fertile plains of the Maasai Mara Wildlife Reserve in East Africa. The seemingly endless landscape stretching off into the distance beneath the bright blue sky is the image that many people associate with the wide-open spaces of the African plains.

is both a buoyancy aid when swimming and provides excellent insulation from cold temperatures.

The *lesser snow goose* (*Chen caerulescens caerulescens*) takes its name from its white plumage, rather than from the snowy areas that it inhabits. They spend the winter along the Gulf of Mexico and in the southern United States, and in spring begin the 12-week, 2,500 mile (4,000 km) journey that takes them to their breeding grounds in the Arctic.

Australian Plain: The Nullarbor Plain

The Nullarbor Plain is a huge dry plateau that stretches some 400 miles (650 km) from Western Australia into South Australia; it is bordered to the north by the Great Victoria Desert and to the south by the Great Australian Bight and covers an overall area of around 100,000 square miles (260,000 sq km).

The Nullarbor Plaint takes its name from the Latin *nullus arbor*, meaning "no tree." The plain's vegetation is mainly blue bush and saltbush, small shrubs that are adapted to the arid, salty environment.

Beneath the surface of the Nullarbor Plain can be found the Nullarbor Caves, a spectacular network of subterranean caves that harbor a variety of unique animals.

Animals of the Nullarbor Plain

The *southern hairy-nosed wombat* (*Lasiorhinus latifrons*) is the smallest of Australia's three species of wombat and it is listed as a threatened species in South Australia where it is the fauna symbol of the region. The Nullarbor Plain is home to the largest population of the southern hairy-nosed wombat.

Like kangaroos, the southern hairy-nosed wombat is a marsupial—a mammal that carries its young in a pouch. They are burrowing creatures and their pouch faces backward to ensure that no soil gets into it as they burrow.

The *naretha bluebonnet parrot (Northiella haematogaster)* is another endangered species of the Nullarbor Plain. They generally grow to around 12 inches (300 mm) in length and have a colorful plumage of reds and blues. Bird smuggling is thought to be a contributing factor to the declining numbers of the naretha bluebonnet parrot.

Previous page: A mother bison *(Bison bison)* and her calf walking on the plains at first light on frosty morning in Yellowstone National Park, Wyoming. The steam from hot spring geysers can be seen rising in the background, partially concealing the treeline in the distance.

Left: An expansive view of shortgrass prairie grassland in Colorado, USA. A lone bison can be seen grazing just below the horizon on the right-hand side.

Below: A family of black-tailed prairie dogs *(Cynomys ludovicianus)* standing above ground, on look out in the Wichita Mountains National Wildlife Refuge, Oklahoma. Their burrows—often referred to as "towns"—can house many thousands of animals. The largest colony ever recorded was in Texas and covered an area of around 25,000 square miles (65,000 sq km) and was home to an estimated 400 million black-tailed prairie dogs.

Top inset: Two female lions *(Panthera leo)* with cubs in the Masai Mara, East Africa. Lions are the only cats that live in groups (known as prides), usually numbering somewhere between 10 and 30 members and comprised of several generations of lionesses and their offspring, along with a smaller number of adult males.

Above left inset: A male cheetah *(Acinonyx jubatus)* standing on a tree in the Maasai Mara, East Africa. The cheetah is the fastest land animal in the world, and can reach speeds of between 50 and 62 miles per hour (80 to 100 kph) over short distances.

Main image: A leopard *(Panthera pardus)* crouching in undergrowth, ready to pounce. Leopards are the smallest of the four big cats—the others being lions, tigers, and jaguars. Mainly nocturnal, they are solitary animals that mainly prey on antelopes and deer, although they are opportunistic hunters and will feed on any animal they can overpower.

Previous page: A large herd of wildebeest (*Connochaetes taurinus*) crossing a fast-flowing river in the Maasai Mara Wildlife Reserve, East Africa during their seasonal migration. The migration instinct is so strong in wildebeest that they will attempt to cross any river in their path—regardless of the depth, current, or even the presence of crocodiles.

Top: A reticulated giraffe (*Giraffa reticulata*)—also known as the Somali giraffe—towering over a small tree in the Samburu Game Reserve, Kenya. Giraffes are the tallest land animal and males (bulls) can reach heights in excess of 18 ft (5.5 m), while the smaller females (cows) grow to around 15 ft (4.5 m).

Above: Two adult white rhinoceros (*Ceratotherium simum*)—also known as the square-lipped rhinoceros—and a calf grazing in Lake Nakuru National Park, Kenya. Unlike the horns of cattle and antelopes, the horns of the white rhinoceros are not made of bone but of keratin fibres.

Left: An elephant (*Elephantidae*) drinking from a waterhole in the Etosha Game Reserve, Namibia. Waterholes provide a permanent water supply during dry season and attract large numbers of animals including herds of zebra, springboks, wildebeest, hartebeest, and giraffes.

Top: A flock of galahs (*Eolophus roseicapilla*)—also known as the roseate cockatoo—flying over the grasslands of the Mitchell Grass Downs in Queensland, Australia. The galah is the most abundant species of cockatoo and their raucous flocks are a common site throughout Australia.

Above: A view of the semiarid shrubland in Mulga Lands, northern New South Wales, Australia during the brief summer flowering season. The yellow flowering cassia shrubs can be seen filling the plains into the distance.

Main image: A summer rainstorm brings much needed moisture to the parched earth of the dry tussock grassland of outback Queensland, Australia. The predominant grass that can be seen on the plain is Mitchell grass, a long lived grass that grows in clay soils as tussocks of two inches (5 cm) to 20 inches (50 cm) in diameter.

Main image: A herd of barren-ground caribou (*Rangifer tarandus groenlandicus*), seen here on their calving grounds on the coastal plain of the Arctic National Wildlife Refuge, Alaska. The barren-ground caribou is a sub-species of caribou, their coat changes color with the seasons—during the summer it is brown and in the winter it becomes much lighter.

Below: An arctic hare (*Lepus arcticus*) gambols among the flowers of the tundra at Hubbart Point, Manitoba, Canada;. Like the caribou, the coat of the arctic hare changes depending on the season. In the winter it is white, in the summer it changes to the blue-grey color seen here—although in the far north the artic hare's coat actually turns almost completely white in the summer.

Below: The Dempster Highway, snaking its way through the vast tundra of Yukon Territory, Canada. In the far distance low-lying clouds, with clear blue skies above, crown the peaks of the Richardson Mountains.

Wetlands & Freshwater

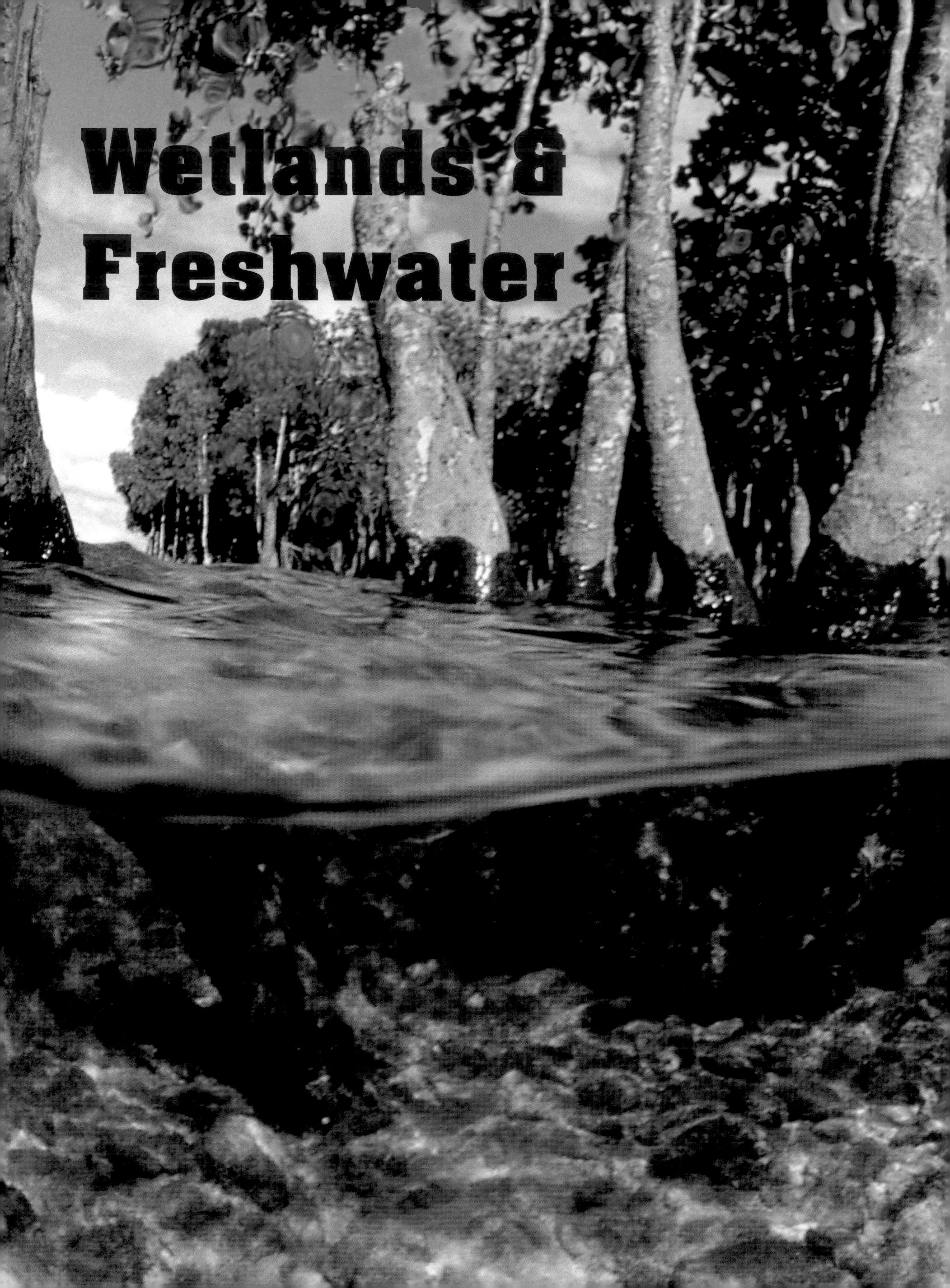

Fresh water is the most valuable natural resource on Earth—without water life cannot exist. Just over 70 percent of our planet is covered by water but most of this is saline sea water—natural fresh water is much rarer, only three percent of the total. Most available fresh water lies in rivers, streams, ponds, and lakes (as well as in underground lakes and reservoirs) from where it is used by people, plants, and animals in everyday life. Luckily fresh water is a renewable resource falling as rain and snow from the atmosphere.

The definition of fresh water is water containing less then 0.5 parts per thousand dissolved salts. However, two thirds of natural fresh water (as opposed to human-cleaned reservoir water) is frozen into the ice caps, glaciers, and ice sheets of the Arctic and Antarctic while much of the rest lies underground in aquifers and subterranean lakes and streams. Only 0.3 percent of freshwater lies on the surface, mostly in lakes, mangroves, and swamp lands and even less, 0.04 percent, is held in the atmosphere. Proportionally, even the great rivers of the world do not contain a significant amount of fresh water. The largest volume of open fresh water is held in lakes such as the Great Lakes of North America, Lake Baikal in Russia, and Lake Victoria in Africa. The Caspian Sea is considered the largest lake in the world because it is landlocked, but its water is saline.

Significant World Rivers

For most people, however, the most often seen fresh water lies in the rivers and streams which dominate the landscape. Rivers flow from inland sources, often many miles away such as meltwater from mountains and glaciers, or run-off from heavy rainfalls. Every continent excepting the Antarctic has its iconic rivers on which many millions of people, flora, and fauna utterly depend. All these large watercourses have stories and legends attached to them and often very direct historic links that have dictated the political future of the peoples who live around them. Some rivers have acted as impassable barriers across which invaders could not transgress, others have acted as highways to the interior for both good and bad.

Even today the length of the longest rivers is disputed: it depends on which tributaries are counted and which watercourses are considered to constitute the same flow. The largest river system in North America is the Mississippi–Missouri–Jefferson which has the third largest catchment area in the world and drains 41 percent of the continental U.S. from part or all of 31 states and two Canadian provinces. The watercourse runs for around 3,900 miles (6,300 km) through ten states (and originally before the river bed moved made up the state boundaries of Iowa, Wisconsin, Illinois, Missouri, Kentucky, Arkansas, Tennessee, and Mississippi), is joined by 19 major rivers and hundreds of smaller tributaries, before draining into the Gulf of Mexico.

In the far Northwest Territories of Canada the huge Mackenzie–Peace–Finlay river system originates in the Great Slave Lake from where it flows 2,635 miles (4,241 km) to the Atlantic Ocean. For five months of the year, between May and October, the river is navigable, but in winter it freezes so hard that it is used as an ice road. In complete contrast the Amazon River in South America is the largest river in the world by volume (and with modern instruments many scientists consider it to be longer than the River Nile, the official longest river) and draws water from the largest drainage basin in the world from the tropical rainforests of Brazil, Peru, Bolivia, Colombia, Ecuador, Venezuela, and Guyana before discharging into the Atlantic Ocean after traveling for approximately 4,300 miles (6,992 km).

The second longest river in South America is the more southerly Paraná which also draws most of its waters from Brazil, but also from Paraguay (it is the border between the two countries), Argentina, Bolivia, and Uruguay for 1,600 miles (3,998 km) to reach the Atlantic Ocean at Rio de la Plata where it discharges with a volume comparable to the Mississippi River delta. Much of its upper length is navigable and serves as an important conduit to the ocean for the inland cities in Paraguay and Argentina.

As befits a large continent, Africa has many large rivers including the famous River Nile, which at 4,135 miles (6,650 km) is officially the longest river in the world and draws is waters from a vast area of Africa including Kenya, Tanzania, Uganda, Rwanda, Congo, Burundi, Ethiopia, Sudan, Eritrea, and Egypt.

On the African west coast the Congo–Chambeshi flows right across west central Africa from its sources in Lake Tanganyika and Lake Mweru in the East African Rift Valley, then through the second largest watershed and second largest rainforest in the world into the Atlantic Ocean beside the town of Muanda. Almost the entire river is navigable.

Although not very long at only 125 miles (200 km) the most important river in the Middle East is the Shatt al-Arab which gathers waters from Iraq, Turkey, Iran, and Syria before discharging into the Persian Gulf. At the southern end, toward the mouth, the river constitutes the border between Iraq and Iran, but control of the waterway has been a cause for bitter war between the pair for centuries.

The Indian subcontinent contains many huge rivers but the biggest is the Indus, which starts with meltwaters from glaciers high up on the Tibetan plateau. The river catches 93 percent of its waters from Pakistan, plus a little from Kashmir and Afghanistan. It is fed by 20 major tributaries and flows into the Arabian Sea at Sindh province after running for some 1,976 miles (3.180 km).

Previous page: View from above and below the waters of a mangrove swamp. These tropical and sub-tropical swamps grow along muddy estuaries of large tropical rivers, and in sheltered intertidal coastal areas such as lagoons, tidal creeks, bays, and inlets. Only about 54 plant species can exclusively cope with the particular demands of life in the mangrove—the freshwater from a river and twice daily inundations of high saline waters of the sea.

Above left: The coastal wetland wilderness of the Bynoe River floodplain near its mouth in the Gulf of Carpentaria, Queensland, Australia. This area of tidal saltflats and mangrove-lined channels provides a rich habitat for an amazing variety of species including estuarine crocodiles, turtles, dolphins, barramundi, and many other forms of fish and marine life.

Right: A comb-crested jacana (*Irediparra gallinacea*) also known as the lotus bird or lily trotter and native of southeast Asia, New Guinea, and the Pacific islands, seems able to walk on water as it trots across the floating leaves of the water lilies in the freshwater floodplain lagoon in the Gulf of Carpentaria, Queensland, Australia. In the wet season, the overflowing waters link up with the flooded grasslands to form Palm Lagoon in the Gilbert-Smithburn river delta and become the largest wetland habitat in Australia. In the hot, dry summer, the much reduced lagoons are life savers for wildlife.

The second biggest river on the subcontinent is the Brahmaputra which also collects its headwaters from water in southwestern Tibet, and then plunges through the Himalaya Mountain range through huge gorges which take it into India. From there it moves into Bangladesh, where it joins the Ganges to create the Sunderbans, the biggest delta in the world on the edge of the Bay of Bengal. Unfortunately the Brahmaputra frequently floods catastrophically following the spring melt thousands of miles away in the Himalayas.

By contrast European rivers are quite modest. The Danube gathers its prime waters in the Black Forest of Germany as well as from much of central Europe: Romania, Hungary, Austria, Serbia, Slovakia, Bulgaria, Bosnia and Herzegovina, Croatia, Ukraine, and Moldova. Its course runs for 1,771 miles (2,850 km) until it creates the Danube delta at the Black Sea in Romania and Ukraine. Historically, the Danube was the eastern border of the Roman Empire for many centuries.

The most famous river in Australia is the Murray at 1,600 miles (2,575 km) long. It rises in the Australian Alps and then runs to form the border between New South Wales and Victoria, turns to flow into South Australia and then into Lake Alexandrina and through to the Indian Ocean at a small river mouth at Goolwa. In conjunction with the Darling River, the Murray drains most of inland Victoria, New South Wales, and south Queensland. The river flow varies considerably depending on the lack of rain, and the Murray has been known to dry up completely on rare occasions.

The third longest river in the world is the Yangtze (also known as Chang Jiang) which rises from glacial meltwaters in Geladandong mountain on the eastern Tibetan plateau. It then runs through Qinghai Province and eventually eastward into the East China Sea at Shanghai. The Han River and four freshwater lakes contribute waters to the Yangtze along its 3,915 miles (6,300 km) long course. Historically, the Chinese have used the Yangtze as the dividing line between north and south China.

The Yellow River (also known as Huang He) is called the "cradle of Chinese civilization" and is the sixth longest river in the world. Its source lies in the Bayan Har Mountains of Qinghai Province in western China from where it runs generally east-west through nine provinces to discharge into the Bohai Sea some 3,389 miles (5,464 km) away. It is notorious for frequent devastating floods, which has led to its other name of "China's Sorrow."

In southeast Asia the mighty Mekong (also known as Lancang Jiang) also rises on the Tibetan plateau before running through Yunnan Province in China and then on through Burma, Thailand, Laos,

Cambodia, and Vietnam for some 2,703 miles (4,350 km) until it reaches the South China Sea. Navigation is notoriously difficult along almost the entire length of the Mekong due to extremes of seasonal flow and the presence of numerous rapids and waterfalls. The flow of the river inverts in Cambodia where the low-tide river level is lower than the high sea tide, which together with the flat landscape makes the Mekong delta particularly prone to flooding even as far up as Phnom Penh.

Freshwater Lakes: North America

Both Canada and the United States contain large bodies of freshwater lakes, the biggest being lakes Michigan and Huron, which together form one hydrological unit. Containing around 2,029 cubic miles (8,458 cu km) of water over a surface area of 45,445 square miles (117,702 sq km), they can be regarded as one body because the channel between them flows in both directions and their surfaces lie at the same elevation. Lake Superior holds a greater volume of water—2,900 cubic miles (12,100 cu km)—but has a smaller surface area of 31,820 square miles (82,414 sq km). Other large bodies of freshwater in North America include Great Bear Lake, Great Slave Lake, Lake Erie, Winnipeg, Ontario, Athabasca, and Reindeer lakes.

Freshwater Lakes: Africa

The numerous lakes of the African Rift Valley are some of the largest, deepest, and oldest lakes in the world, though not all of them contain fresh water. Of the eight lakes of the Kenyan Rift Valley only lakes Baringo and Naivasha are freshwater, the others are soda lakes. The largest central African body of water is Lake Victoria on the borders of Tanzania, Kenya, and Uganda. It is the second largest freshwater lake and the largest tropical lake in the world at 26,560 square miles (68,800 sq km). Situated in an elevated plateau at the western end of the Rift Valley, Victoria collects water from a vast 71,040 square mile (184,000 sq km) catchment area and is the primary source of the White Nile that drains northward to join the greater Nile, which—after a journey of some 4,000 miles (6,437 km)—empties into the Mediterranean sea. It is relatively shallow and geological examinations have shown that Lake Victoria has dried up three times in its lifetime (last time 17,300 years ago). The other huge Rift Valley lakes are (from north to south), Albert (empties into the White Nile), Edward (empties into the White Nile), Kivu (empties into the Congo River), Tanganyika (the second largest in the world by volume also empties into the Congo River) and Malawi (alkaline, it drains into the Shire River and ultimately the Zambezi).

Freshwater Lakes: Australia

Australia is the driest continent of all, though it does contain numerous salt lakes. One of these, Lake Eyre is the largest ephemeral lake in the world and also the largest nominally freshwater lake in Australia—on the rare occasions when it fills, which is about four times a century. It is located at the lowest point in Australia at -49 ft (-15 m) below sea

Above left: The Colorado River in Glenn Canyon, in the Page Area of Arizona. The river is essential to the lives and livelihoods of the American southwest with its waters being used for drinking, municipal use, animal drinking water, and agricultural use—particularly crop irrigation for which almost 90 percent is taken—plus many other uses.

Right: Hugely dramatic in any season, Niagara Falls is the most powerful waterfall in North America and lies on the border between the U.S.A. and Canada. When the flow is at its peak, more than six million cubic feet (168,000 cu m) of fresh water tumble over the crest every minute, though the average flow is a mere four million cubic feet (110,000 cu m) a minute.

level in the remote deserts of northern South Australia. Some water remains in isolated lakes between the salt pans during the dry season, but when the monsoons arrive in the Queensland outback Lake Eyre starts to fill, as it partially does anyway with local heavy rains. Every three years or so a five foot (1.5 m) flood occurs, once in ten years it will be a 13 ft (4 m) flood and the lake will partially fill. However, the lake seems only to become really big in a La Niña year (most recently 1974–76). Despite the way it fills, the water is clear enough for freshwater fish such as bream to survive. But as the 18 in (450 mm) salt pan crust dissolves in the water over a period of some six months, the salinity increases so much that most, if not all, of the fish die. By the end of the following summer the water in the lake will have almost completely evaporated.

Freshwater Lakes: South America

By volume the largest lake in South America is Lake Titicaca located high in the Andes mountains on the border between Peru and Bolivia. It is the highest commercially navigable lake in the world at 12,507 ft (3,812 m). Titicaca is filled by rain and via five major river systems—Ramis, Coata, Ilave, Huancané, and Suchez—which in turn are fed by meltwater from glaciers. It loses water through strong winds and evaporation in the intense summer sunlight and and only about ten percent drains out of a single outlet to Lake Poopó, making Titicaca technically an almost closed lake.

Lake Nicaragua was called La Mar Dulce (the Sweet Sea) by the Spanish conquistadors, because the water is pure and so vast they thought it a sea although it is only 43 ft (13 m) at its deepest. In fact it is part of the largest drainage basin in South America and is fed by rain and numerous rivers.

Freshwater Lakes: Russia

Situated in southeast Siberia close to the border with Mongolia, 25 million year old Lake Baikal is the oldest (formed during the Palaeozoic, Mesozoic, and Cenozoic periods) and deepest at 5,577 ft (1,700 m) in the world; it is also the seventh largest lake in the world. It contains 20 percent of the world's total of unfrozen fresh water in its 12, 162 square miles (31,500 sq km). Lake Baikol has exceptionally clear water with up to 131 ft (40 m)visibility in which upwelling currents and vertical water movements oxygenate the deeper waters.

Due to its age and unusually fertile depths, Lake Baikal has the richest freshwater fauna anywhere. Surrounded by a wall of mountains—the Khamar Daban mountains in the south, the Primorskiy and Baikalskiy ranges to the west, and the Barguzinskiy and Ulan-Bagasy ranges to the east—Lake Baikal is fed by 335 main tributaries which flow from the mountains, principally the rivers Selenga, Turka, Barguzin, and Upper Angara.

The largest lake in Europe is also in Russia and also a freshwater lake. Lake Ladoga is situated in the Republic of Karellia and Leningrad Oblast in northwestern Russia. It has a surface area of 6,834 square miles (17,700 sq km) most of which comes in via tributaries, the principal rivers being the Volkhov, Svir, and Vuoksa. Lake Ladoga drains out through the Neva River into the Gulf of Finland. Once rich with 48 different types of fish the

Above right: Xiling Gorge on the Yangtse River in China. Deceptively peaceful in this view, before it was made relatively safe for navigation in 1950 the entire length of the gorge was swirling with whirlpools as it rushed through seven small gorges and two stretches of fierce rapids. Even now the gorge is notorious for its many dangerous reefs as it zigzags for about 49 miles (79 km). The gorge's spectacular beauty is threatened by the Three Gorges Dam project.

Below: Only fully appreciated from space, this vast alluvial fan radiates outward across the desolate landscape from the feet of the Kunlun and Altun mountain ranges on the southern border of the Taklimakan Desert in XinJiang Province, China. The river shows bright blue while the old riverbeds show up as white. Alluvial fans are caused by perennial freshwater springs which are fed by mountain snow melt.

numbers have been drastically reduced by commercial fishing. Between November and December the lake freezes around the edge and then in the center between January and March.

Freshwater Lakes: China

The largest freshwater lake in China is Poyang in Jiangxi Province and home to over half a million migratory birds in season. Recently sand has been dredged from the lake and the muddying waters threaten to lead to the demise of the finless porpoise (*Neophocaena phocaeniodes*). Dredging makes it difficult for the animals' sonar to work in the muddied waters, so they can't easily find fish to eat, avoid large objects, or dodge away from shipping lanes busy with a constant stream of vessels moving in and out to the Yangtse River.

The Poyang Lake Bird Protection Area claims to be the largest bird sanctuary in the world and features many rare species of migrating birds in November, which arrive from Siberia, Mongolia, Japan, North Korea, and the colder regions of China: they start to leave when the warmer weather begins in March. In winter 95 percent of the world's population of Siberian cranes (*Grus leucogeranus*) arrive from Arctic Russia and western Siberia to escape the bitter cold of their homelands.

Lake Dongting in northeastern Hunan Province is the second largest freshwater lake in China. It is a large body of shallow water surrounded by mountains. The lake acts as a flood basin for the Yangtse River which flows into it between July and September, so its size depends on the season. The climate here is warm and humid though cold winds sometimes blow in from the mountains to cool the subtropical air. Its most famous and endangered resident is the finless porpoise whose existence is threatened, as at Lake Poyang, by sand dredging and the huge number of ships passing through the waters.

Freshwater Lakes: Antarctica

As the remote continent of Antarctica is explored and studied it has become apparent that huge underground lakes are hidden deep under the 1.8 mile (3 km) deep ice cap. About 140 subglacial lakes have been discovered so far and the largest of these is Lake Vostok, discovered in 1996, 13,000 ft (4,000 m) below the surface under Russia's Vostok Station. Lake Vostok is 155 miles (250 km) long by 31 miles (50 km) at its widest point and contains an estimated volume of 1,300 cubic miles (5,400 cu km) of fresh water. Analysis has shown that the water there averages a million years old, sits at a temperature of 26.6° F (-3° C), and remains liquid despite being below freezing point due to the extreme pressure from the weight of the ice above and geothermal heat from the Earth's interior. Ice core samples indicate that the lake supports life in the form of ancient bacteria. Research has also shown that the lake has tides and rises between 0.3 - 0-7 in (1 - 2 cm) depending on the position of the sun and moon.

Above: Red River gums (*Eucalyptus camaldulensis*) reflected in the winter flood waters of the Murrumbidgee River of New South Wales, Australia. Found along the watercourses of much of mainland Australia, this species of eucalyptus seems untroubled by periodic flooding, can grow up to 148 ft (45 m) tall and may well live to be a thousand years old.

Left: The darling lily (*Crinum flaccidum*)—also known as the Murry lily—only flowers after wet weather. It is seen here in the flood waters of the Paroo River, eastern Australia. Originating near Quilpie in western Queensland after heavy summer rains, the Paroo River meanders south through a network of swamps, waterholes, and flood plains until it eventually spreads into the vast floodplains of western New South Wales. The Paroo usually evaporates in the Australian heat before it reaches the River Darling.

Main image: View of the channels, sandbars, deltaic islands, intertidal flats, beaches, and mangroves of the Burdekin wetland complex in north Queensland, Australia. The entire delta region is important for both commercial and recreational fishing. Probably thanks to frequent floods and cyclones, the watercourses and coastlines have changed considerably over the last 40 years or so.

Above left: A wattled jacana (*Jacana jacana*) walking across lily pads in Pantanal. Named for the red wattles down the side of its yellow bill, this common wetland bird lives among the floating vegetation eating insects and other invertebrates. Most unusually, each female mates with several males, one of which goes on to incubate and then look after the chicks.

Above: Another Pantanal resident, the capped heron (*Pilherodius pileatus*) waits patiently and statue-like for its prey—fish, insects, and frogs. It is a commonly found bird across tropical South America in rivers, swamps and freshwater lakes, particularly in the Orinoco and Amazon basins.

Above left: An aerial view of Pantanal, the world's largest wetland. This UNESCO World Heritage Site lies principally in the Brazilian states of Mato Grosso and Mato Grosso do Sul. In the wet season the floodwaters submerge over 80 percent of the area, in turn giving rise to the greatest diversity of aquatic plants in the world. Many experts believe that this region—some 75,000 square miles (195,000 sq km)—contains the most dense flora and fauna ecosystem in the world.

Left: Lurking in the rushing waters of a river in the Panatal lie a line of Yacare caiman (*Caiman yacare*). These alligators are native to the rivers, lakes, and wetlands of northern Argentina, southern Brazil, southern Bolivia, and Paraguay. Reaching lengths of almost 10 ft (3 m) these fearsome looking animals live off a diet of fish, aquatic invertebrates—especially snails—and snakes when they can catch them.

Right: The Pantanal has its origins in the Paraguay River but swells with an average annual rainfall of around 55 in (1,400 mm). Between December and May (the rainy season) the water generally rises by a further 10 ft (3 m) nourishing the land it covers, almost all of which is used for agriculture and ranching.

Above: A stormy day on the Indus river plain, west of Leh, Ladakh, in the Himalayas. The River Indus rises in Tibet and then flows west into north India before turning south into Pakistan and then on through into the Arabian Sea.

Main image: A satellite photograph of the Ganges River delta where it empties into the Bay of Bengal. The rivers Ganges, Brahmaputra, and Meghna together make up the largest delta in the world most of which comprises an impenetrable swamp and mangrove forest called the Sunderbans. Becoming a UNESCO World Heritage site in 1997, the Sundarbans covers 1,645 square miles (4,262 sq km) in India and a further 2,315 square miles (6,000 sq km) in Bangladesh.

Above: A typical coconut palm fringed backwater river in the evening light near the village of Kumarakom, in Kottayam district, Kerala, south India. This popular tourist destination is well known for the Kumarakom Bird Sanctuary which covers over 14 acres (5.5 ha) and was established following a government initiative to protect local wildlife.

Above: Two Indian or Bengal tigers (*Panthera tigris*) enjoying the water in Bandhavgarh National Park. Here, they live in the tropical mangrove swamps, but their habitat and numbers are severely threatened. Tigers are protected in the Sundarbans Tiger Reserve which was created in 1973. At a little short of 1,000 square miles (2,585 sq km) it is the largest tiger reserve and national park in India.

Left: An osprey (*Pandion haliaetus*) rising from the waters with a fish caught in its claws in Kangasala, Finland. Living on an exclusive diet of fish, the osprey always nests near water. It is a big raptor with a wingspan that can reach a length of 6 ft (1.8 m). The osprey enjoys a worldwide distribution across temperate and tropical regions and is found on all the continents except Antarctica.

Below left: An aerial view of the rivers and coastal marshes of north Norfolk, England—an ecosystem of outstanding importance. The entire coastal edge encompasses some of the most important salt marsh, dunes, intertidal flats, shingle, and grazing marsh in Europe, but the whole environment is a constantly shifting coastline of unique scientific and ecological value.

Right: A kingfisher (*Alcedo atthis*) in flight with a fish held securely in its beak. Found besides lakes, canals, streams, dykes, and canals extensively throughout the freshwater areas of Britain—with the exception of Scotland—and eastern and southern central Europe, the kingfisher feeds exclusively on aquatic animals. In winter, when ice covers the water, it moves to tidal marshes and the shoreline.

Below: A young European otter (*Lutra lutra*) spotted chewing fish on the rocky shoreline of Sutherland, Scotland. Otters are opportunist feeders but will only live in unpolluted freshwaters so long as there is a constant supply of birds, frogs, insects, crustaceans, and the occasional small mammal to make a meal. Otters will live beside the coast but still require unpolluted fresh water nearby where they can clean their fur.

Main image: Lesser flamingoes (*Phoenicopterus minor*) gather in their thousands at Lake Nakuru in central Kenya. This freshwater lake, which formed millennia ago within the East African Rift Valley, is the ancestral home to millions of flamingoes who get their pink coloration from the beta carotene in the brine shrimps that provide the bulk of their diet.

Below inset: A hippopotamus (*Hippopotamus amphibus*) mother protectively watches her newborn calf. Although bulky, hippos can run remarkably fast over short distances—30 mph (48 kph)—and are very agile in the water. Hippos are easily provoked to attack and are indisputably credited with being the most dangerous animal species in Africa.

Main image: Pa-hay-okee Overlook in Everglades National Park, Florida, is a 12 ft (3.6 m) observation platform at Shark River Basin and provides sweeping views of the sawgrass "river of grass." The area is accessed by a boardwalk which allows visitors to traverse through a hardwood hammock, then a bald cypress forest: Pa-hay-okee means "grassy waters" in the local Seminole language.

Top: The top predator of the Everglades is the alligator (*Alligator mississippiensis*), seen here in Sanibel Island, a small barrier island just off the southwest coast of Lee County, Florida. The American 'gator is a native of the wetlands and groves of the southeastern U.S. Although it is dangerous, this alligator is essential to the well being of the Everglades as it helps to control the populations of rodents and other animals which would otherwise overwhelm the marshlands.

Above: Everglades National Park is a wetland habitat and home to the red mangrove (*Rhizophora mangle*). This shrub supports itself above the brakish water on stilt roots and has been used by Native Americans as a remedy for a wide variety of ailments: asthma, boils, wounds, angina, fever, jaundice, leprosy, and many other illnesses.

Next page above: A West Indian manatee (*Trichechus manatus*) mother and calf. Manatees are very agile in water and have been hunted for centuries—almost to the point of extinction—but do not possess avoidance behavior as they evolved without natural predators. They live on a diet of sea grass and other vegetation, plus the occasional fish and invertebrate.

Next page below: A shoal of mangrove snappers (*Lutjanus griseus*) in Three Sisters Springs, on the Crystal River in Florida. These snappers live in mangroves, tidal creeks, and grass beds. Mature fish weigh up to 10 lb (4.5 kg) and are a popular target for fishermen because they are a challenge to catch and very good to eat.

Credits

2 Ecoscene/Jon Bower
6—7 Ecoscene/Steven Kazlowski
8—9 Ecoscene/Christine Chelmick
10 Ecoscene/Ian Harwood
11 Ecoscene/Joel Creed
12 *left* Ecoscene/Peter Cairns
12 *above* Ecoscene/Chinch Gryniewicz
13 Ecoscne/Fritz Polking
14 Ecoscene/Anthony Cooper
15 Ecoscene/Paul Kennedy
16 Mike Norton/Fotolia
17 *above* Rob/Fotolia
17 *left* Intro Venture/Fotolia
18—19 Ecoscene/Pauline Horton
20 *top* Ecoscene/Steven Kazlowski
20 —21 Ecoscene/Steven Kazlowski
22 Ecoscene/Graham Neden
23 jqz/Fotolia
24 *main image* Ecoscene/Pearl Bucknall
24 Ecoscene/Peter Cairns
25 *top* Ecoscene/Peter Cairns
25 *above*/Ecoscene/Frank Blackburn
26 Ecoscene/Edward Bent
27 *above top* Ecoscene Chinch Gryniewicz
27 *below* Ecoscene Luc Hosten
27 *above* Ecoscene/Fritz Polking
28—29 Paul Hampton/Fotolia
30 *left* Ecoscene/John Corbett
30—31 below Ecoscene Kirill Grekov
31 *above* Ecoscene Fritz Polking
32—33 *above* Dimitry Pichuqin/Fotolia
32 *far left* Ecoscene/Chinch Gryniewicz
33 Granitepeaker/Fotlia
34—35 *main* Ecoscene/Robert Nichol
34 *far left* Ecoscene/John Farmer
34 *center* Ecoscene/Nick Hanna
35 *below* Ecoscene/Nick Hanna
36—37 Ecoscene/Phillip Colla
38 Ecoscene/Michael Gore
39 Ecoscene/NUR
40 Ecoscene/Visual & Written
41 Ecoscene/Richard Dirscherl
42 Ecoscene/Phillip Colla
43 Ecoscene/Richard Dirscherl
44 Ecoscene/Phillip Colla
45 Ecoscene/Steven Kazlowski
46 *above* Ecoscene Reinhard Dirscherl
46 *left* Ecoscene/Jeff Collett
47 Ecoscene/Reinhard Dirscherl
48 Ecoscene/John Lewis
49 *above* Ecoscene/John Lewis
49 *right* Ecoscene Kjell Sandved
50—51 Ecoscene/Reinhard Dirscherl
52—53 *main image* Ecoscene/Reinhard Dirscherl
52 *below left* Ecoscene/Reinhard Dirscherl
53 *below middle* Ecoscene/Reinhard Dirscherl
54 *below right* Ecoscene Peter Tatton
54—55 i-Stockphoto.com/Dejan Sarman
56 *above* Ecoscene/Reinhard Discherl
56 *right* Ecoscene/ Phillip Colla
57 Ecoscene/Philip Colla
58—59 Ecoscene/Christine Osborne
60—61 Ecoscene/Andrew Brown
62 Ecoscene/Fritz Polking
63 Ecoscene/Sally Morgan
64 Ecoscene/ John Corbett
65 Ecoscene/Fritz Polking
66—67 Ecoscene/Mike Buxton
68 *above* Shutterstock/Pichugin Dmitry
68 *inset* Ecoscene/Sally Morgan
69 *top right* Shutterstock/Pichugin Dmitry/ 69 right Ecoscene/Gehan de Silva
70 *main image and far left below* Ecoscene/Wayne Lawler
70 *below* middle Image Science and Analysis Laboratory, NASA-Johnson Space Centre. 'The Gateway to Astronaut Photography of Earth'
71 *below* Ecoscene/Wayne Lawler
72—73 main image Ecoscene/Reinhard Dirscherl
73 *above* Ecoscene John Corbett
74 Image courtesy of USGS, National Center for EROS and NASA Landsat Project Science Office
75 Image courtesy of USGS, National Center for EROS and NASA Landsat Project Science Office
76—77 Ecoscene/Frank Blackburn
78—79 Ecoscene/Fritz Polking
80 Ecoscene/Andrew Brown
81 Ecoscene/Wayne Lawler
82 Ecoscene/Andrew Brown
83 Ecoscene/Sally Morgan
84 Ecoscene/Andrew Brown
85 *top* Ecoscene/Fritz Polking
85 *above* Ecoscene/Steven Kazlowski
86 *above* Ecoscene/Paul Thompson
86 *below* Ecoscene/Peter Cairns
87 *right* Ecoscene/Erik Schaffer
87 *inset* Ecoscene/Andrew Brown
88—89 Ecoscene/Stuart Baines
90—91 *main image* Ecoscene/Frank Blackburn
90 *insets* Ecoscene/Peter Cairns
91 *inset left* Ecoscene/Jack Milchanowski
91 *inset right* Ecoscene/Hugh Clark
92—93 *main image* Ecoscene/Andrew Brown
93 Ecoscene/Sally Morgan
94 Ecoscene/Wayne Lawler
94 Ecoscene/Edward Bent
95 Ecoscene/Wayne Lawler
95 Ecoscene/Michael Maconachie
96—97 Ecoscene/Fritz Polking
98—99 i-Stockphoto.com/Warwick Lister-Kaye/
100 Ecoscene/Robert Weight
101 Ecoscene/Steven Kazlowski
102 Ecoscene/Visual & Written
103 Ecoscene/Genevieve Leaper
104 Ecoscene/Stuart Donachie
105 Ecoscene/Robert Weight
106 *main image* Ecoscene/Fritz Polking
106 insets Ecoscene/Steven Kazlowski
107 Ecoscene/Steven Kazlowski
108—109 Ecoscene/Steven Kazlowski
110—111 Ecoscene/Steven Kazlowski
112—113 Ecoscene/Fritz Polking
114 Ecoscene/Michael Gore
115 *above* Ecoscene/Fritz Polking
115 *right* Ecoscene/Graham Neden
116—117 Ecoscene/Fritz Polking
118 Ecoscene/John Farmar
119 *above* iStockphoto.com/Alexander Hafemann
119 *right* Ecoscene/Fritz Polking
120—121 Ecoscene/Graham Neden
122—123 Ecoscene/Sally Morgan
124 Ecoscene/Wayne Lawler
125 Ecoscene/David Wootton
126 Ecoscene/Robert Gill
127 Ecoscene/Chinch Gryniewicz
128 Ecoscene/Karl Ammann
129 Ecoscene/Edward Bent
130—131 Ecoscene/Frtitz Polking
132 *top left* i-Stockphoto.com/Morley Read
132 *middle left* i-Stockphoto.com/Steve Geer
132 *below left* i-Stockphoto.com/Mark Kostich
132 *below* i-Stockphoto.com/Anneke Schram
133 *above* i-Stockphoto.com/Oliva Falvey
133 *right* i-Stockphoto.com/Miguel Cesar
134 Ecoscene/Kjell Sandved
135 *above* Ecoscene/Michael Gore
135 *below* Ecoscene/Andrew Brown
136 Ecoscene/Fritz Polking
137 *above* Ecoscene/Fritz Polking
137 *right* Ecoscene/Bryan Knox
138—139 Ecoscene/Wayne Lawler
140 Ecoscene/Colin Conway
141 *above* Ecoscene/Wayne Lawler
141 *right* inset Ecoscene/Paul Kennedy
141 *right* Ecoscene/Paul Kennedy
142—143 *main image* Ecoscene/Wayne Lawler
142 *above* Fotolia/Daniel Budiman
143 *top* Ecoscene/Wayne Lawler
143 *above* Ecoscene/Robert Gill
144 *left* Ecoscene/Alastair Shay
144 *below* Ecoscene/Sally Morgan
145 Ecoscene/Kjell Sandved
146—147 Ecoscene/Fritz Polking
148 Ecoscene/Andrew Brown
149 Ecoscene/Peter—153 Ecoscene/Fritz Polking
154—155 Ecoscene/Peter Cairns
156 Ecoscene/Andrew Brown
157 *below* Ecoscene/Jack Milchanowski
158—159 Ecoscene/Fritz Polking
160—161 Ecoscene/Fritz Polking
162—163 Ecoscene/Tony Wilson-Bligh
163 *top* Ecoscene/Robert Gill
163 *above* Ecoscene/Fritz Polking
164—165 Ecoscene/Wayne Lawler
166—167 *main image* Ecoscene/Steven Kazlowski
166 Ecoscene/Visual & Written
167 Ecoscene/Andy Binns
168—169 Ecoscene/Reinhard Dirscherl
170 Ecoscene/Wayne Lawler
171 Ecoscene/Wayne Lawler
172 Ecoscene/Fritz Polking
173 Ecoscenen/Graham Kitching
174 image courtesy of USGS National Center for EROS and NASA Landsat Project Science Office
175 Ecoscene/Stephen Coyne
176—177 Ecoscene/Wayne Lawler
178—179 Ecoscene Fritz Polking
180—181image courtesy of USGS National Center for EROS and NASA Landsat Project Science Office
180 *left inset* Ecoscene/Chinch Gryniewicz
180 *middle inset* Ecoscene/Chinch Gryniewicz
181 *right inset* Ecoscene Satyendra Tiwari
182 *above* Ecoscene/Fritz Polking
182 Ecoscene/David Wootton
183 *right* Ecoscene/Fritz Polking
183 *below* Ecoscene/Peter Cairns
184—185 *main image* Ecoscene/Michael Gore
184 *inset* Ecoscene/Kjell Sandved
186 *main image* Ecoscene/Andrew Brown
187 *above* Ecoscene/Andrew Brown
187 *top* Ecoscene/Fritz Polking
188 Ecoscene/Phillip Colla

Index

Abyssopelagic Zone 39
Aconcagua, mountain 15
Adelie penguin 115
Adirondack Mountains, New York 11
Afghanistan 171
Africa 10, 14, 27, 31, 53, 63, 68, 80, 85, 90, 124, 126, 127, 129, 151, 158, 163, 170, 171, 173, 184
African Plate 10
African rainforest 127, 128
African Rift Valley 173
African savannah elephant 152, 163
Ala Shan Desert 65
Alaska 20, 25, 38, 45, 80, 81, 85, 100, 104, 107, 110, 111, 152, 166
Albert, Lake 173
alder tree 79
Alexandrina, Lake 172
Algeria 63
alligator 179, 187
Alps (European) 10, 13, 14, 31, 86
Alpine ibex 14
Altai Mountains 12, 31, 65
Altiplano 22
Altun, mountains 174
Amazon, basin 179
Amazon Rainforest 10, 127
Amazon River 127, 128, 171
Amboseli National Park, Kenya 31
American badger 91
American bison 149, 150, 151, 157
American Museum of Natural History 65
Amur, basin 83
Andaman Sea 42
Andes 10, 12, 13, 15, 22, 62, 73, 82, 127, 174
Angara, river 174
Angel, James "Jimmie" 135
Angel Falls, Venezuela 135
Angola 63
Antarctic 13, 43, 62, 78, 100, 101, 102, 103, 104, 105, 115, 117, 119, 129, 148, 170, 175, 183
Antarctic fur seal 103
Antarctic Ice Sheet 101, 104
Antarctic Peninsula 102, 105, 119
Antarctic Plate 10
Antarctic skua 103
Antarctic snow petrel 103
antelope 14, 151, 152, 159, 163
Antelope Valley Poppy Reserve, California 60
Appalachian Mountains 110
Arabia 129
Arabian Sea 171, 180
Araucariaceae 80
Arctic 11, 45, 78, 83, 85, 100, 101, 102, 104, 107, 110, 111, 115, 152, 153, 170
Arctic Circle 100, 101, 107
Arctic fox 101, 110
Arctic ground squirrel 102, 111
Arctic grouse 101
Arctic hare 101, 166
Arctic National Wildlife Refuge, 107, 111, 166
Arctic Ocean 38, 43, 107, 115
Argentina 13, 62, 82, 119, 129, 171, 179
ash tree 80
Asia 10, 12, 15, 31, 33, 65, 79, 82, 83, 90, 91, 124, 128, 171
Asian rainforest 128
Asiatic wild ass 150
aspen tree 79, 85
Assal, Lake 72
Atacama Desert 62, 65, 73
Athabasca, Lake 173
Atlantic Ocean 13, 38, 42, 43, 45, 56, 63, 85, 127, 128, 171
Atlantic spotted dolphin 42
Atlas Mountains 10, 63
Australia 10, 15, 41, 49, 56, 64, 71, 80, 83, 94, 95, 124, 129, 138, 139, 142, 153, 165, 171, 172, 173, 177
Australian Alps 172
Australian giant cuttlefish 49
Australian Plain 153
Australian Plate 10
Australian rainforest 129
Austria 27, 172
Azilal 10
Azores 42

bactrian camel 31
Badwater Basin 61
Baffle Creek 138
Baikal, Lake 170, 174
Baikalskiy, range 174
bald eagle 12
Bali 141
Balinese macaque 141
bamboo lemur 137
Bandhavgarh National Park 181
Bangladesh 172, 180
Barguzin, river 174
Barguzinskiy, range 174
Baringo, Lake 173
barramuni 171
basswood tree 80
Bathypelagic Zone 39
Bavarian Forest National Park 79, 82
Bay of Bengal 172
Bayan Har Mountains 172
Bear Island 107
beech tree 79, 80, 83
Bengal tiger 181
Bering Sea, Alaska 45
bighorn sheep 12, 85
birch tree 79, 80, 85
bitternut hickory 83
Black Hills, South Dokota 11
black bear 85
Black Forest 172
black oak 83
black orchid 144
Black Sea 31, 172
black-tailed prairie dog 150, 157
blue grama 149
blue morpho butterfly 132
blue whale 41, 56
bobcat 85
Bohai Sea 172
Bohemian Forest, Czech Republic 79, 82
Bolivia 13, 127, 133, 171, 174, 179
Booth Island 119
Borneo 56, 141
Bosnia 172
Botswana 68
bottle tree 80
Brahmaputra, river 172, 180
brain coral 49
Brazil 85, 127, 129, 133, 171, 179
broadleaf southern beech 83
Bromo, Mount 35
Brooks Range 20, 101, 107
brown bear 15, 20
buffalo 14
buffalo grass 149
buff-tailed coronet 135
Bukit Barisan Selatan National Park 128
Bulgaria 172
Burchella zebra 152
burdekin plum tree 139
Burdekin Wetland 177
Burma 172
Burundi 14, 171
Buttercup, Glacier 102
Bynoe River 171

Cacusus Mountain Range 31
Cairngorms National Park, Scotland 24, 25, 90
Callabonna, Lake 15, 35
Cambodia 173
Cameroon 128
camphor 14
Canada 11, 25, 80, 85, 91, 100, 110, 111, 124, 148, 149, 152, 166, 167, 171, 173
Canadian Shield 149
canopy layer 125, 129
Cape Renard 119
capped heron 179
caribou 25, 101, 105, 107, 111, 152, 166
Caspian Sea 31, 170
cats claw creeper 139
Catskills, New York 11
Caucasus Mountains 31
cedar 14
Central African Republic 128
Chad 63
Challenger Deep 42
Chambeshi, river 171
cheetah 158
Chihuahuan Desert 60, 62, 65
Chile 12, 62, 82
chimpanzee 137
China 33, 65, 68, 79, 83, 96, 97, 172, 174, 175
chit palm 135
Churchill River 85
clouded leopard 143
Coata, river 174
coconut palm 181
Colombia 12, 127, 133, 135, 171
Colorado River 61, 173
Columbia Plateau 60
comb-crested jacana 171
condor, Andean 13
Congo 171
Congo basin 127
Congo peacock 128
Congo River 128, 171, 173
conifer 78, 79, 80, 86
Cook, Mount 35
coral reefs 41
Cordillera de la Costa 62
cork oak 80, 83
Cotopaxi, volcano 12, 13
cottonwood tree 62
crabeater seal 103
Croatia 172
crown jellyfish 53
Crystal River 187
cymbidium orchid 144

Daintree Rainforest 129
dall sheep 20
Danakil Desert 72
dancing doll orchid 144
Danube, River 172
darling lily 177
Darling, River 172, 177
Darwin Island 56
Dawson – Lambton Glacier 119
Death Valley 61
Death Valley National Monument Park 60
Democratic Republic of the Congo 14, 128
Dempster Highway 167
dendrobium orchid 144
Denmark 25
Diamantina, river 64
Doi Inthanon National Park 144
dolphin 171
Dome mountains 11
Dongting, Lake 175
dormouse 91
Douglas fir 80, 81
Dyer Island, South Africa 56

Ecuador 12, 127, 135, 171
East African mountains 14
East African olive tree14
East African Rift Valley 171, 184
East China Sea 172
Ediacara Hills 15
Edward, Lake 173
Egypt 63, 171
Elbrus, Mount 31
elephant 14, 31, 142, 163
elephant seal 103
elm tree 79
emergent layer125, 129
emperor penguin 103, 119
England 79, 82, 86, 90, 183
Epipelagic Zone 38, 39
Equatorial Guinea 128
Erebus, Mount 102
Erg Chebbi, Merzouga 60
Erie, Lake 173
Eritrea 171
estuarine crocodile 171
Ethiopia 72, 171
Etosha Game Reserve 163

eucalyptus tree 83, 95, 177
Eurasian Plate 10
Europe 10, 13, 25, 31, 79, 80, 82, 83, 86, 90, 91, 124, 172, 174, 183
European otter 183
Everest, Mount 15, 33, 39
Everglades National Park 187
Exit Glacier, Alaska 20
Eyre, Lake 64, 173, 174

Fee Glacier 27
Filchner Ice Shelf 101, 105
Finke River 64
Finland 80, 86, 100, 174, 183
Finlay River 171
finless porpoise 175
Fiordland National Park, New Zealand 35
Firehole River 149
firs 80, 85
Flinders Range 15, 35
Flinders Ranges National Park 15
forest floor 126
France 10, 27
Fraser Island World Heritage Area 124
French Guiana 127
freshwater lake 173, 174, 175

Gabon 128
galah 165
Galapagos fur seal 41
Galapagos Islands 41, 56
Ganges, River 172, 180
gazelle 151
Geladandong, mountain 172
gerenuk 152
Germany 82, 172
giant panda 79
giant redwood 93
Gilbert, river 171
glacier 105
glacier buttercup 102
Glacier Point 17
glasswing butterfly 126
Gobi Desert 12, 65, 68
golden eagle 12, 25, 85
Gombe Stream National Park 137
Gondwana Rainforest 129
gorilla 128
Graian Alps 10
Grand Teton National Park 17
Great Australian Bight 153
Great Barrier Reef 41, 49
Great Basin Desert 60, 62
Great Bear Lake 173
Great Lakes, North America 170
Great Plains, Africa 151
Great Plains, North America 148
Great Slave Lake 171, 173
Great Victoria Desert 64, 153
great white shark 56
Greater Khingan Range 65
green sea turtle 56
Greenland 25, 100, 104
Greenland Ice Sheet 104
grizzly bear 12, 111

Grumeti Controled Area 151
Guatemala 135
Gulf of Carpentaria 171
Gulf of Mexico 61, 171
gum tree 94
Gunnison prairie dog 150
Gunung Leuser National Park 128, 141
Guyana 127, 171

Hadalpelagic Zone 39
Half Dome 17
Han River 172
Hangayn, mountains 65
Hawaii 38, 45
hemlock 80, 81
Herzegovina 172
Heysen Range, Australia 35
hickory 80, 83
Himalaya Mountains 10, 12, 15, 33, 97, 150, 172, 180
hippopotamus 184
hoary marmot 20
Hoh Forest, Washington 82
holm oak 83
Hooker Glacier 35
Hopen 107
Huancané, river 174
Huangshan Mountains 96
Hudson Bay, Canada 85
Humboldt Current 62
Humboldt Lake, Nevada 60
hummingbird 135
humpback whale 38, 45
Hungary 172
Huon pine 83
Huron, Lake 173
hyrax 27

ice sheet 104
iceberg 104, 119
Iceland 100
Iguacu National Park 129
Iguazu Falls 129
Ikorongo Controled Area 151
Ilave, river 174
India 43, 129, 150, 172, 180, 181
Indian Ocean 38, 42, 43, 49, 53, 56, 128, 172
Indonesia 35, 43, 46, 141, 142
Indus, River 171, 180
Industrial Revolution 79
International Hydrographic Organization 38, 43
International Union for the Conservation of Nature 56
Inuit 100
Iran 171
Iraq 171
Ireland 82
Italy 10, 11, 27

Japan 39, 80, 175
Jasper National Park 15, 20
Java Trench 42
Jefferson River 170

jellyfish 39
Joshua Tree National Park 62

Kalahari Desert 68
kangaroo 153
kapok tree 126
Karri Forest, Australia 94
Kashmir 171
Katmai National Park, Alaska 20
Kazakhstan 80
Kenai Fjords National Park 20
Kenya 14, 31, 151, 152, 163, 171, 173, 184
Kenyan Rift Valley 173
Kerinci Seblat National Park 128
Key Summit 35
Khamar Daban, mountains 174
Kibo 14
Kilimanjaro, Mount 14, 31
killer whale 115
Kinabalu, Mount 129
King Billy pine 83
kingfisher 183
Kivu, Lake 173
koala 95
Koussi, Mount 63
Kumarakom Bird Sanctuary 181
Kunlun, mountains 174

Lagoda, Lake 174
Lake Nakuru National Park 163
Laos 172
larch tree 79, 80
leafcutter ant 132
leafy seadragon 49
Lemair Channel 119
leopard 14, 159
leopard seal 103, 115
lesser flamingo 184
lesser snow goose 153
Libya 63, 74
lion 158, 159
lionfish 53
lithosphere 10
Little Rann of Kutch National Park 150
Loa River 62
Loliondo Controled Area 151
Los Glaciares National Park 119
lynx, European 12

Maasai Mara 149, 152, 158
Maasai Mara National Reserve 151
Maasai Mara Wildlife Reserve 152, 163
MacDonnell Ranges 64
Machu Picchu 22
Mackenzie River 149, 171
Madagascar 129, 137
Malawi, Lake 173
Malayan sun bear 143
Malaysia 124, 129
Maldives 43
Mali 63
Maligne, Lake 20
mangrove snappers 187

Manitoba Tundra 102
maple tree 79, 80
Mara River 152
Marakech 10
Mariana Trench 39, 42
marmot 14, 85
Maswa Game Reserve 151
Matterhorn, the 10
maues marmoset 127
Mauretania 63
Mawenzi 14
McMurdo Sound 102
Mediterranean Sea 13, 63, 79, 80, 173
meerkat 68
Meghna, river 180
Mekong River 172, 173
Mens, the, Sussex 90
Mesopelagic Zone 38, 39
Mexican prairie dog 150
Mexico 56, 91, 126, 133
Michigan, Lake 173
Mid-Atlantic Range 43
Middle Atlas Mountains 10
Milwaukee Deep 42
Mississippi River 170, 171
Missouri River 170
Mitchell Grass Downs 165
Mojave Desert 60, 61, 62
Moldova 172
Mongolia 31, 65, 68, 174, 175
Mongolian steppes 150
monkey 14
monkey puzzle tree 80, 82
Mont Blanc 10, 13
Mont Blanc Massif 10
moose 85
Morning Glory Pool 17
Morocco 10, 63
mountain gorilla 27, 137
mountain lion 85
Mulga Lands 165
Mulligan, river 64
Murmansk 100
Murray River 172
Murrumbidgee River 177
musk ox 101, 105, 107
Mweru, Lake 171

Naivasha, Lake 173
Nakuru, Lake 184
Namib – Naukluft National Park 75
Namib Desert 62, 63, 68, 75
Namibia 63, 68, 75, 163
naretha bluebonnet parrot 153
Nenets 100
Nepal 33
Neumeyer Channel 101
Neva, river 174
New Caledonia 80
New Guinea 42, 142, 171
New Jersey 10
New Mexico 11
New Zealand 15, 35, 80, 83, 124
Nganasan 100
Ngorongoro Conservation Area 151
Niagra Falls 129, 173
Nicaragua, Lake 174
Niger 63